T0344406

Numerical Calculations for Process Engineering Using Excel VBA

Numerical Calculations for Process Engineering Using Excel VBA provides numerical treatment of process engineering problems with VBA programming and Excel spreadsheets. The problems are solving material and energy balances, optimising reactors and modelling multiple-factor processes. The book includes both basic and advanced codes for numerical calculations. The basic methods are presented in different variations tailored to particular applications. Some macros are combined with each other to solve engineering problems. Examples include combining the bisection method and binary search to optimise an implicit correlation, combining golden section search with Euler's method to optimise a reactor and combining bisection code and Euler's method to solve steady-state heat distribution. The text also includes nonconventional examples such as harmony search and network analysis. The examples include solutions to common engineering problems such as adiabatic flame temperature, plug flow reactor conversion, batch reactor, heat diffusion and pinch analysis of heat exchanger networks. The VBA code is presented with mathematical equations and flowcharts, enabling the audience to adopt the solutions to different problems. The book contains many demonstrations of numerical techniques to guide users. It also includes useful summaries of VBA commands/functions and Excel-predefined functions accessible in VBA.

While the book is developed primarily for undergraduate students, the book is a helpful resource for postgraduate students and engineers.

Numerical Calculations for Process Engineering Using Excel VBA

Chi M. Phan

CRC Press
Taylor & Francis Group
Boca Raton London New York

CRC Press is an imprint of the
Taylor & Francis Group, an **informa** business

Designed cover image: © Shutterstock

First edition published 2024
by CRC Press
2385 NW Executive Center Drive, Suite 320, Boca Raton FL 33431

and by CRC Press
4 Park Square, Milton Park, Abingdon, Oxon, OX14 4RN

CRC Press is an imprint of Taylor & Francis Group, LLC

© 2024 Chi M. Phan

Reasonable efforts have been made to publish reliable data and information, but the author and publisher cannot assume responsibility for the validity of all materials or the consequences of their use. The authors and publishers have attempted to trace the copyright holders of all material reproduced in this publication and apologize to copyright holders if permission to publish in this form has not been obtained. If any copyright material has not been acknowledged please write and let us know so we may rectify in any future reprint.

Except as permitted under U.S. Copyright Law, no part of this book may be reprinted, reproduced, transmitted, or utilized in any form by any electronic, mechanical, or other means, now known or hereafter invented, including photocopying, microfilming, and recording, or in any information storage or retrieval system, without written permission from the publishers.

For permission to photocopy or use material electronically from this work, access www.copyright.com or contact the Copyright Clearance Center, Inc. (CCC), 222 Rosewood Drive, Danvers, MA 01923, 978-750-8400. For works that are not available on CCC please contact mpkbookspermissions@tandf.co.uk

Trademark notice: Product or corporate names may be trademarks or registered trademarks and are used only for identification and explanation without intent to infringe.

ISBN: 978-1-032-42828-4 (hbk)
ISBN: 978-1-032-42829-1 (pbk)
ISBN: 978-1-003-36451-1 (ebk)

DOI: 10.1201/9781003364511

Typeset in Times
by Apex CoVantage, LLC

Contents

Preface

I came across Excel VBA as a fresh engineering graduate in 2000. For an engineer with a strong software background, VBA seems too primitive. Yet the software allows a powerful connection with databases in Excel spreadsheets and effective communication with others. Over the next two decades, Excel VBA became an essential part of my research and teaching. It was a pleasure to explore and create new calculation methods for new and not-too-new problems. Occasionally, I review my previous code and see how it can be improved by twisting it around. While there are many guides to writing simple VBA for a particular numerical method, combining this simple code can open unlimited opportunities to solve a complex problem. Such combinations take the coder's vision beyond computer programming and mathematical algorithms. This exciting creativity was the primary motivation for this book. I hope the readers enjoy the journey, too.

I want to thank Dr Gordon Ingram, Dr Anita Hyde and Dr Cuong Nguyen for their insightful discussion on the art of numerical methods. Finally, I thank Quy Ngo, Safin Phan and Calleigh Phan for their unconditional support during the book's development.

About the Author

**Chi M. Phan** (PhD, BEng) has taught in the Department of Chemical Engineering at Curtin University for over 15 years. He has taught classes in process analysis, process calculations, process modelling and simulations. In addition to Curtin University, he taught at the University of Newcastle (Australia) and the University of Hyogo (Japan). He has developed many numerical models for his research in surface science.

1 Introduction to Excel and VBA

1.1 MICROSOFT EXCEL

Microsoft Excel is a spreadsheet program written and distributed by Microsoft for computers. It features an intuitive interface and capable calculation and graphing tools, which have made Excel one of the most popular microcomputer applications. It is overwhelmingly the dominant spreadsheet application available. All Microsoft Office versions (including Apple Mac OS versions) currently have Excel with Visual Basic applications.

1.1.1 VISUAL BASIC FOR APPLICATIONS

Since the 1990s, Excel has included Visual Basic for Applications (VBA), a programming language based on Visual Basic, which adds the ability to automate tasks in Excel and to provide user-defined procedures for use in worksheets. The procedure is attached to the particular Excel file in which it was developed and can only run on that Excel file. Excel workbooks of this type have to be saved as macro enabled (.xlsm) in the latest Microsoft Excel versions. Standard Excel files, without macros, can be saved as. xls files. It is necessary to save as. xlsm after VBA code is created. All attached macros will be removed by wrongly saving the file as. xls. These macros can only run within the file if multiple Excel files are opened simultaneously. Macros can be copied and pasted to different Excel files if needed. VBA macros can also

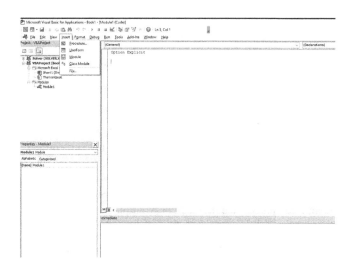

FIGURE 1.1   VBA editor window.

DOI: 10.1201/9781003364511-1

be created by recording. Macro recording can produce VBA code replicating user actions, thus allowing simple automation of regular tasks. However, writing, editing and debugging macros in the VBA editor is more effective for numerical calculations. This editor screen can be accessed by hitting the keystroke **Alt + F11**.

By default, new Excel workbooks have no module. The users must insert the modules. Once inserted, the module can be renamed. All macros in different modules of the same Excel workbook can be used in any spreadsheet.

VBA implements Microsoft's Visual Basic, built into all Microsoft Office applications (including Apple Mac OS versions). VBA is closely related to Visual Basic but can only run code from within a host application rather than as a stand-alone application. As a result, VBA can be used to control one application from another application (for example, automatically creating a Word report from Excel data).

1.1.2 Objects in Excel

VBA manipulates objects in Excel. The objects can be workbooks, worksheets, cell ranges and charts. Objects are arranged in a hierarchy, and one object can act as a container for other objects. At the top of the object hierarchy is Excel, which is an object called Application, and it can contain other objects such as workbook objects and Command Bar objects. For example, the workbook object contains objects such as worksheet objects and chart objects. A worksheet object, in turn, contains Range objects.

Objects of the same type form a collection. For example, the **Worksheet** collection consists of all the worksheets in a particular workbook. You can refer to an object by specifying its position in the object hierarchy, using a dot as a separator. For example, you refer to Range A1 in worksheet Sheet1 in workbook Book1.xls as

```
Application.Workbooks("Book1.xls").Worksheets("Sheet1").
Range("A1")
```

If you omit specific references, Excel uses the active objects. For example, if Book1 and Sheet1 are active, then the preceding reference can be simplified to:

```
Range("A1")
```

Objects have properties. For example, a Range object has such properties as Value and Address. A Chart object has such properties as Title and Type. You can use VBA to determine object properties and to change properties.

The user can refer to a property of an object by combining the object name with the property name, separated by a period. For example, you can refer to the value in cell A1 on Sheet1 by

```
Worksheets("Sheet1").Range("A1").Value
```

You can assign values to variables. A variable is a named element that stores data. You can use variables to store such things as values, text or property settings. To assign the value in cell A1 on Sheet1 to a variable called X_value, use the following statement:

```
X _ value = Worksheets("Sheet1").Range("A1").Value
```

In addition to properties, objects have methods. A method is an action Excel performs with an object. For example, one of the methods for a Range object is **Clear-Contents**. This method clears the contents of the range. You specify a method by combining the object with the method, separated by a dot. For example, the following statement clears the contents of cell A1:

```
Worksheets("Sheet1").Range("A1").ClearContents
```

While VBA code can manipulate the cells in an Excel spreadsheet directly, it is not always used in numerical calculations. Instead, a macro is often used for the purpose. The purpose and limitations of macros are explained in the next section.

1.1.3 SUB AND FUNCTION PROCEDURES IN VBA

In Excel, the users can perform actions in VBA by writing (or recording) code in a VBA module. The users can also view and edit existing VBA modules in the Visual Basic Editor (VBE). In Microsoft Excel's VBE, users can create two types of procedures: sub and function. Both are sets of commands that are used to perform specific tasks in VBA. A sub, also known as a subroutine or sub-procedure, is a code used to perform a specific task, but it does not return any value. On the other hand, a function, also known as a user-defined function procedure, executes a specific task determined by the Excel user and returns a result. Within the function, the result is stored as the name of the function. The function can take cell values in an Excel spreadsheet as arguments. VBA subs and functions can include all the constructs of modern programming languages, including arrays, branches and loops.

Many macros can be written into the same module. The program automatically separates them at the end of each macro, with either End Sub or End Function. Codes 1.1 and 1.2 show examples of sub and function. The first function, named **double1**, can be used as a standard Excel built-in function, that is, enter "=double1(cell location)" in any spreadsheet cell. The cell location can be changed to any other cell ("=double1(A1)" or a single numerical value ("=double1(103)"). The second function, named **double2**, calculates the double value of a specific cell, cell A1, in the active worksheet.

Code 1.1. Two VBA functions for doubling a number.

```
Function double1(x)
double1 = x * 2
End Function
Function double2()
Dim x
x = Worksheets("Sheet1").Range("A1").Value
double2 = x * 2
End Function
```

VBA subs are more useful for "action" procedures (e.g. formatting cells, displacing value in Message Box). To use a sub to export results to an Excel spreadsheet, a set value can be used. However, the range of cells is specified within the sub and cannot be moved as with functions. One effective way to use the sub in an Excel spreadsheet is to create a shape and assign a sub to it. Then the user can click on the shape. Alternatively, the sub can be assigned to a particular keystroke. Using the Excel menu and keyboard commands, subs can be created by recording. However, the recorded macros cannot incorporate logic, branching and looping.

Code 1.2. Two VBA subroutines for doubling a number.

```
Sub double3()
Dim xx
xx = Worksheets("Sheet1").Range("A1").Value * 2
MsgBox "The answer is "& xx
End Sub
Sub double4()
Dim x
x = Worksheets("Sheet1").Range("A1").Value * 2
Worksheets("Sheet1").Range("A2").Value = x
End Sub
```

To use **double3** or **double4** in Code 1.2, one can create a Command Button in an Excel spreadsheet and then assign the subroutine to it (Figure 1.2). Subsequently, the user needs to click on the message box to close it. The subroutines do not take the input value as an argument. In contrast, functions allow both (i) a numerical value as an argument and (ii) the value stored in a specific cell as the argument.

The output from the subroutine cannot be used by other VBA functions or subroutines. This disadvantage of subroutines limits their usage in numerical calculations.

As seen, both subs and functions can do the same calculations. The two essential function features are the return of a value (or a set of values) and multiple uses in the spreadsheet. Once created, functions are available in the Excel library. VBA functions work in the same way as formulas and built-in functions in Excel. A function

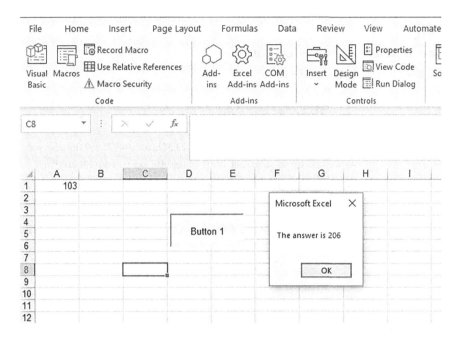

FIGURE 1.2   Using a subroutine in Excel spreadsheet.

can be used multiple times within the same Excel workbook. Users can create custom functions for any action and then access the functions from a cell or a direct reference from a cell. In addition, the function return value can be combined with another sub/function. For this reason, VBA functions are more convenient for numerical calculations and will be used mainly in this book.

1.2 KEY CHARACTERISTICS OF VBA PROGRAMMING

1.2.1 VARIABLES IN VBA

For numerical calculations, it is essential to declare variables. VBA variables can be a single value or a set of values, as in an array or a matrix. The Dim statement can specify the variable type within a VBA function by typing "As . . .".

```
Dim x As Integer
Dim y1,y2 as string
Dim A(2,10) as long
Dim B() As double
```

If omitted, the default type of variable is **Variant**, which means the variable can be any value (text, numbers, dates, time or objects). For numerical types of variables, VBA has predefined ranges of values (as in Table 1.1). Therefore, it is important to

TABLE 1.1

Range of Values for Numerical Data Types

Data Type	Range of Values
Byte	0 to 255
Integer	–2,147,483,648 to 2,147,483,647
Short	–32,768 through 32,767 (short integer)
Long	–9,223,372,036,854,775,808 through 9,223,372,036,854,775,807 (long integer)
Single	-3.402823×10^{38} to $-1.401298 \times 10^{-45}$ for negative values, 1.401298×10^{-45} to 3.402823×10^{38} for positive values
Double	$-1.79769313486232 \times 10^{308}$ to $-4.94065645841247 \times 10^{-324}$ for negative values, $4.94065645841247 \times 10^{-324}$ to $1.79769313486232 \times 10^{308}$ for positive values
Decimal	+/–79,228,162,514,264,337,593,543,950,335 for no decimal points, +/–7.922816 2514264337593543950335 for 28 places to the right of the decimal

consider the expected range of the value. If any calculated value within the code is outside the defined range, the VBA code will give an error (#VALUE) without specifying the problem.

The ranges might change with the Excel version. The specified values can be checked in detail on the Microsoft website.

https://docs.microsoft.com/en-us/dotnet/visual-basic/language-reference/data-types/

Code 1.3. Size of a matrix (since the function name is Variant, the code can return a text string or a numerical value).

```
Function SM(A)
'return the size of the matrix
Dim N as integer
N = A.Rows.Count
If A.Columns.Count <> N Then
    SM = "Not a square matrix":
Else
    SM = N
End If
End Function
```

1.2.2 STATIC AND DYNAMIC ARRAYS

An array is a set of variables. Excel/VBA allows a multi-dimensional array with up to 60 dimensions. If omitted, the array variable is **Variant**, and elements within the array can be any type. The elements don't have to be the same types in this case.

Array variables can be either static or dynamic. A dynamic array is an array that can be resized, as opposed to a static array, which cannot be resized. For static array, the size is fixed at declaration, with the statement **Dim**. Dynamic data can be reinitiated by the statement **Redim**. An existing array can be extended while keeping the existing value using the **ReDim Preserve** statement.

ReDim can be used to redefine the variable type as well. However, it would create confusion for code users. More often, **ReDim** is used to redefine the dimension of an array or matrix. It is constructive for output with a variable number of answers.

The values of an array can also be reset with **Erase** instead of using **Redim**. While **Redim** resizes the dimension of the array, the **Erase** statement reinitialises the elements of fixed-size arrays and releases dynamic-array storage space. The initialisation depends on the type of variables. For example, Erase resets the fixed numeric array to zero and the fixed variant to Empty.

1.2.3 RETURNING MULTIPLE VALUES IN A SPREADSHEET

Excel allows users to call VBA functions to spreadsheets in matrix format. Unlike other software, in which the output dimension is defined (for example, precisely the same as the input), Excel/VBA allows a flexible output.

For example, if the function gives the answers as a 10×10 matrix, that is a two-dimensional array. If the user can take a 2×2–range Excel spreadsheet and enter the function, the output will be the first four elements of the answers. If the user selects an 11×11 range and enters the function, all the extra cells will have a default value of zero. It is important to note that **ReDim** cannot be used with the function's name. Consequently, a practical way to manage multiple outputs is to declare a dynamic variable, for example, **output**(), and re-dimension as needed. After calculations, the desired outcomes are allocated to elements of **output**. As the output is variant, its element can be of any data type independently. For example, output(1,1) can be text, and output(1,2) is numerical. If an array is used, the outcomes are in a horizontal range of cells. Finally, the function name can be assigned to an array. Hence, size and elements are included. The output array should have one or two dimensions only. Only the first indexes are used if the array has more than two dimensions (remember, VBA allows up to 60 dimensions!).

The standard way to select a range of cells is to type in the function in the formula bar and then hit the keystroke **Ctrl+Shift+Enter**. The bracket {} is automatically included. If the user selects a range larger than the output, all extra cells will display "%NA". This fixed-size result is also known as the LEGACY ARRAY formula.

In some later Excel versions, the user can use the SPILLED ARRAY formula. In these versions, the user only needs to select one cell (this corresponds to the output array's first element) and ensure that other cells (below and to the right) are empty. The data will automatically spill over the required range. This is similar to the spilling behaviour of a dynamic array. The advantage of this method is that you don't need to know the dimension of the data matrix. If the neighbouring cells are not empty, Excel will return **#SPILL!** error.

Code 1.4. Output a matrix in an Excel spreadsheet.

```
Function mymatrix(n)
Dim MyArray()
ReDim MyArray(1 To 10, 1 To n)
mymatrix = MyArray
End Function
Function mymatrix2(n)
Dim MyArray()
ReDim MyArray(1 To 10, 1 To n, 1 To 20)
mymatrix2 = MyArray
End Function
Function mymatrix3(n)
Dim MyArray()
ReDim MyArray(1 To 10, 1 To n, 1 To 20)
mymatrix3 = MyArray(1, 1, 2)
End Function
```

The previous presents three functions of the array. Using function mymatrix(5) in the latest Excel version will return a 5×10 table of zero. Typing mymatrix2(5) will return the #VALUE (VBA cannot express a three-dimensional matrix in an Excel spreadsheet). Typing mymatrix5(5) will return a single value of zero. If the VBA function returns a two-dimensional matrix, another VBA function can use that output.

1.2.4 For . . . Next Loop Structure

The numerical method is solving mathematical or physical equations using computers. In other words, numerical methods solve complicated problems by using many repetitive steps of simple calculations. Computer software code is developed to serve this purpose, that is, making computers repeat simple calculations. As a result, the loop structure is an essential part of most numerical methods.

VBA has two types of iterative loops: **For Next** and **Do . . . While** structures. For Next loops have two options: **For Each Next** and **For Next**. These structures are useful when the number of repetitions is predetermined. For example, in the **For Each Next** loop, the iteration loops through a collection of items. The collection can be a collection of objects or a list of numbers. Examples of collections of objects can be cells in a range, worksheets in a workbook and any other object you interact with within Excel. In the case of cells, the order of the loops is left to right, then down (starting in the first cell in the range and looping to the next column in the same row, then going to the next row).

Example of For Each Loop: the periodic table is tabulated in an Excel file. The table has four properties for each element: atomic number, atomic symbol, English element name and atomic mass. Write a VBA to find the molecular weight of a symbol.

Since the element symbols are not in one-dimensional rows or columns, the For Each Loop is needed. The function in code (Code 1.5) can be used in any spreadsheet

TABLE 1.2
Periodic Table (Short Version) in Excel Spreadsheet

	A	B	C	D	E	F	G	H	I	J	K	L	M	N	O
1	1														
2	H														
3	hydrogen														
4	1.008														
5	3	4											5	6	7
6	Li	Be											B	C	N
7	lithium	beryllium											boron	carbon	nitrogen
8	6.968	9.012											10.81	12.01	14.01
9	11	12											13	14	15
10	Na	Mg											Al	Si	P
11	sodium	magnesium											aluminium	silicon	phosphorus
12	22.99	24.31											26.98	28.09	30.97
13	19	20	21	22	23	24	25	26	27	28	29	30	31	32	33
14	K	Ca	Sc	Ti	V	Cr	Mn	Fe	Co	Ni	Cu	Zn	Ga	Ge	As
15	potassium	calcium	scandium	titanium	vanadium	chromium	manganese	iron	cobalt	nickel	copper	zinc	gallium	germanium	arsenic
16	39.10	40.08	44.96	47.87	50.94	52.00	54.94	55.85	58.93	58.69	63.55	65.38	69.72	72.63	74.92
17	37	38	39	40	41	42	43	44	45	46	47	48	49	50	51
18	Rb	Sr	Y	Zr	Nb	Mo	Tc	Ru	Rh	Pd	Ag	Cd	In	Sn	Sb
19	rubidium	strontium	yttrium	zirconium	niobium	molybdenum	technetium	ruthenium	rhodium	palladium	silver	cadmium	indium	tin	antimony
20	85.47	87.62	88.91	91.22	92.91	95.95	98	101.1	102.9	106.4	107.9	112.4	114.8	118.7	121.8
21	55	56	71	72	73	74	75	76	77	78	79	80	81	82	83
22	Cs	Ba	Lu	Hf	Ta	W	Re	Os	Ir	Pt	Au	Hg	Tl	Pb	Bi
23	cesium	barium	lutetium	hafnium	tantalum	tungsten	rhenium	osmium	iridium	platinum	gold	mercury	thallium	lead	bismuth
24	132.9	137.3	175.0	178.5	180.9	183.8	186.2	190.2	192.2	195.1	197.0	200.6	204.4	207.2	209.0

within the Excel workbook. The code can be used in other Excel work in the same folder by using the following:

Workbooks("example1–1.xlsm"). Worksheets("PeriodicTable"). Range("C1:BP46")

Code 1.5. Example of For Each Next loops to calculate the molecular weight.

```
Function MW(name)
Dim Cell As Range
MW = "Element name is not found"
For Each Cell In Worksheets("PeriodicTable").
Range("C1:BP46")
    If StrComp(name, Cell.Value) = 0 Then
          MW = Cell.Offset(2, 0). Value 'MW is stored
2 cells below the Symbol
    End If
Next Cell
End Function
```

The For Next loop follows a sequence of numbers (the data is in Table 1.2). The default step (1) is used if the step is omitted. Alternatively, any integer (including negative values) can be used. The For Next loop is more useful for numerical calculations and will be used frequently in this book.

1.2.5 Do . . . Loop Structures

The second type of VBA repetition loop is the **Do** loop structure. The critical difference between **Do** . . . loops and the **For Next** loop is that the termination condition (and the number of repeated iterations) is determined within the loop. There are four options for Do loops, as presented in Table 1.3. The conditions are adjusted to the loop structures. In addition, VBA allows the **While . . . Wend** structure (structure 5 in Table 1.3), which is equivalent to the **Do While . . .** loop.

TABLE 1.3
Different Do Loop Formats

Loop format	Description	Example
Do While . . . Loop	Runs 0 or more times while the condition is true	Do While "condition is met" Loop
Do . . . Loop While	Runs 1 or more times while the condition is true	Do . . . Loop While "condition is met"
Do Until . . . Loop	Runs 0 or more times until the condition is true	Do Until "condition is met" Loop
Do . . . Loop Until	Runs 1 or more times until the condition is true	Do . . . Loop Until "condition is met"
While . . . Wend	Same as Do While . . . loop	

Code 1.5 shows two loop structures that are used to find the smallest number in the Fibonacci series which is larger than x.

Code 1.6. Examples of Do While ... loop and Do ... Loop Until.

```
Function LFn(x)
Dim f0, f1, f2
f0 = 0: f1 = 1
Do While f2 < x
    f2 = f0 + f1
    f0 = f1: f1 = f2
Loop
LFn = f2
End Function
Function LFn _ 2(x) 'Excel don't accept LFn2
Dim f0, f1, f2
f0 = 0: f1 = 1
Do
    f2 = f0 + f1
    f0 = f1: f1 = f2
Loop Until f2 > x
LFn _ 2 = f2
End Function
```

A common problem with the loop is the requirement for initial conditions. For example, the numerical method requires code to repeat the iterations, so the calculated error is smaller than 0.01. The calculated error is updated within the loop. One can use the condition as "**calculated_error** > 0.01" with structure 1 or 2. Alternatively, one can use "**calculated_error** < 0.01" and structure 3 or 4. Either way, the initial value of **calculated_error** must be initialised before the loop.

1.2.6   OPTIONS IN VBA CODE

1.2.6.1   Option Base

The VBA default base of an array and matrix is 0, meaning the first element's index is zero. This might cause confusion when doing numerical coding, in which the standard practice uses an index from 1. However, this can be overcome by using Option Base 1.

1.2.6.2   Option Explicit

Using Option Explicit forces the coder to declare all their variables. Using Option Explicit at the start of your Excel VBA code is strongly recommended. The option will avoid typos in variables, which are difficult to find visually. A typical mistake is mistyping "xl" as "x1".

The option can be turned on automatically for all newly created VBA modules by selecting "Require variable declaration" in VBA Options (Figure 1.3).

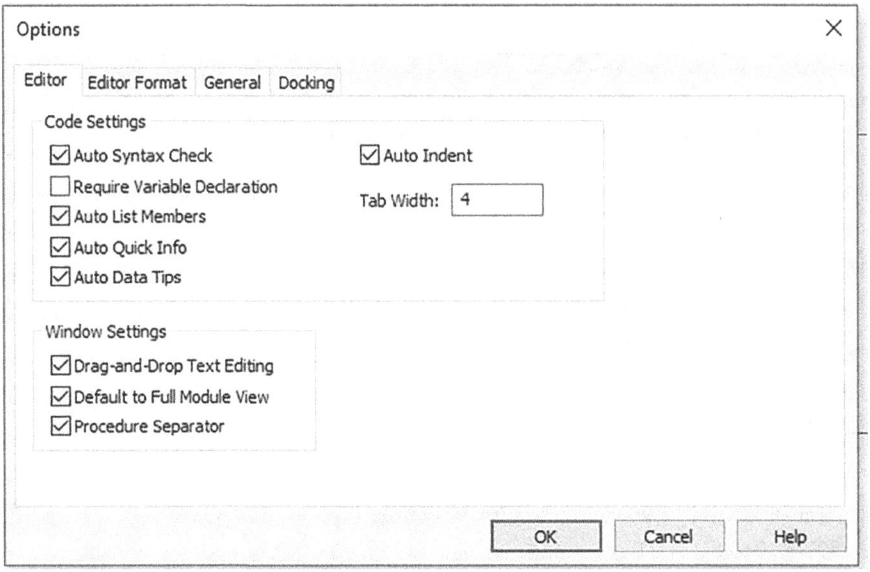

FIGURE 1.3 Screenshot of options in VBA.

1.2.7 GET, SET OR CHANGE SPREADSHEET CELL VALUE

A sub or a function can obtain the value of the cells. The previous example can be rewritten so that the sub will export the output to a specific cell on an active spreadsheet. Hence, the cell's value can be set or changed (if it is not empty). This can be done using the Range or Cells object.

```
Range("A2").Value=1
Cells(2,1).Value =1
```

It should be noted that the first index in Cells is the row number, and the column number is the second index (1 corresponds to column A). For example, Cells(10,2) corresponds to B10, and Cells(2,10) corresponds to cell J2. To get the value from cell A2 of the active spreadsheet, one can use the following:

```
x = Range("A2").Value
x = Cells(2,1).Value
```

The command is useful for selecting values as inputs for a numerical procedure. Alternatively, one includes the cells as arguments for the function.

> **Code 1.7. Calculate the average value of three numbers stored in cells A1, A2 and A3, and export the value to cell B1.**
>
> ```
> Sub ave _ 3()
> Dim ave
> ave = (Range("a1").Value + Range("a2").Value +
> Range("a3").Value) / 3
> Range("B1").Value = ave
> End Sub
>
>
> Function ave _ 4()
> Dim ave
> ave = (Range("a1").Value + Range("a2").Value +
> Range("a3").Value) / 3
> Range("B1").Value = ave
> 'Worksheets("Sheet1").Range("B1").Value = ave
> ave _ 4 = ave
> End Function
>
>
> Sub ave _ 5()
> Dim anything
> anything = ave _ 4()
> End Sub
> ```

Please note that the function **ave_4** does not work by being directly called from an Excel spreadsheet, that is, using it as a user-defined function. The reason is how Excel handles VBA functions. When Excel uses a user-defined function, the function can only access the ranges handed to it via parameters. Any access to other ranges will result in a circular reference and potential cancellation of the execution. However, if a function is called within a subroutine, Excel does not mind. Hence, using sub **ave_5** in combination with function **ave_4** will generate the same result as sub **ave_3**.

1.2.8 Spreadsheet Functions

Some, but not all, worksheet functions in Excel can be called to use in VBA using the **WorkSheetFunction** object. The full list can be found on the Microsoft website: https://docs.microsoft.com/en-us/office/vba/excel/concepts/events-worksheetfunctions-shapes/list-of-worksheet-functions-available-to-visual-basic.

The appendix of this book also contains a list of useful functions (mostly numerical) and their explanations. Some of the useful functions are discussed here. Visual Basic functions do not use the **WorksheetFunction** qualifier. Therefore, a VBA function may have the same name as a Microsoft Excel function and work differently. For example, **Application.WorksheetFunction.Log** and **Log** will return different values. In addition, Excel and VBA have an extended list of text-related functions not included in this book.

1.2.8.1   Pi()

Excel has the function Pi(), which returns the value of Pi accurately to 15 digits. The function can be called in VBA.

```
Dim pi
pi = WorksheetFunction.pi()
```

1.2.8.2   Sum, Average, Min, Max

These functions return the operation on a set of values. The values can be included individually or entered as an array.

```
x _ aver= WorksheetFunction.average(a(1), a(3), (a(5))
x _ sum = WorksheetFunction.sum(x())
```

1.2.8.3   Round, Roundup, Rounddown

Round, Roundup and **Rounddown** return a number to the nearest decimal. These functions require two arguments, the value and the position of the decimal place. The position can be negative or positive. If the position is zero, the function returns the nearest (either up or down) integer. The ROUND function rounds a number up or down, depending on the value of the last digit. Examples of **round, roundup** and **rounddown** are shown in the following.

```
x= WorksheetFunction.roundup (12.01,0)' return 13
y= WorksheetFunction.rounddown(14.91, 1) 'return 14.9
z= WorksheetFunction.round(101.91,-1)' return 100
w= WorksheetFunction.round(101.91,1)' return 101.9
```

1.2.8.4   Ln, Log and Log10

VBA has its own Log function, which returns the natural logarithm of a number. Excel has three logarithm functions: LN, LOG and LOG10. LN and LOG10 require one argument, and the base is fixed (at the natural base and 10, respectively). Spreadsheet function LOG requires two arguments, in which the second argument is the base of the logarithm.

Within VBA is a function LOG, which requires only one argument. This VBA LOG function is the same as **WorksheetFunction.LN**. To evaluate the logarithm of any base without using a spreadsheet function, one can use the indirect methods in Code 1.8.

Code 1.8. Two examples of logarithms in VBA.

```
Function log _ anybase(x,base)
log _ anybase= log(x)/log(base)
End Function
Function log _ any2(x,base)
log _ any2= WorksheetFunction.log(x,n)
End Function
```

1.2.8.5   Tanh, Cosh, Sinh, Atanh, Acosh, Asinh

These are hyperbolic and inverse hyperbolic functions and can be used in standard mathematic formulae. Excel and VBA have normal trigonometric functions (the sine, the cosine, the tangent and their reciprocals).

1.3   AVOIDING ERRORS IN VBA PROCEDURES

1.3.1   ERRORS

Making errors is part of programming. For Excel/VBA, errors can come in many forms (Table 1.4). Finding errors can be frustrating and take much work. For numerical methods, sometimes the error does not come from the coding but from the values of inputs.

Two common errors in using VBA are **#REF!** and **#VALUE!**. The **#REF!** error can happen when the VBA function name has some letter followed by a number, for example, PF2. This can be easily avoided by adding a special symbol, as in PF_2. The **#VALUE!** error can happen when the function has invalid calculations, such as being divided by zero, taking the square root of a negative number or having a calculated value outside the range of a predefined variable. The source of the error can be tricky to find when it arises from the secondary function.

One way to prevent errors is to use **IsError**, **IsNull** or the Excel spreadsheet functions **IsNA** and **IsNumber**. A list of these functions and their operation can be found in the appendix.

TABLE 1.4
Errors When Using VBA Functions in Excel Spreadsheets

Error	Meaning
#NAME?	This error is displayed when Excel does not recognise text in a formula. For example, the name of a function may need to be spelled correctly.
#N/A	Excel displays this error when a value is unavailable to a function or formula.
#NULL!	A space was used in formulas referencing multiple ranges; a comma separates range references.
#NUM!	A formula has invalid numeric data for the type of operation (including VBA functions).
#REF!	A reference is invalid.
#VALUE!	The wrong type of operation or function argument is used.

Code 1.9. Avoiding dividing by text input or zero.

```
Function dividing(x)
If WorksheetFunction.IsNumber(x) Then
    If x <> 0 Then
    dividing = 1 / x
    Else
    dividing = "input is zero"
    End If
Else
dividing = "input is not a number"
End If
End Function
```

1.3.2 ENDLESS LOOP

This is the worst problem for coding. Since most numerical methods employ a repeating loop, it is quite a common problem. The first approach is to use a **For Next** loop instead of a **While** loop. When it is necessary to have a **While** loop, a limitation on the number of iterations should be used. Examples of loops are shown in Code 1.10. These two codes will return 89 for any x greater than 55. The answers are incorrect (as with the functions in Code 1.7), but the code will stop. In this book, the technique is applied often. It is a simple and effective method to avoid endless loops. It is noteworthy that the two combined conditions use different logical functions (And and Or).

Code 1.10. Two While loops with a maximum number of iterations.

```
Function LFn _ 3(x)
Dim f0, f1, f2, i
f0 = 0: f1 = 1
Do While f2 < x And i < 10
f2 = f0 + f1
f0 = f1: f1 = f2
i = i + 1
Loop
LFn _ 3 = f2
End Function

Function LFn _ 4(x)
Dim f0, f1, f2, i
f0 = 0: f1 = 1
Do
f2 = f0 + f1
f0 = f1: f1 = f2
i = i + 1
```

```
Loop Until f2 > x Or i > 10
LFn _ 4 = f2
End Function
```

1.3.3   DISPLACING IMMEDIATE VALUE

The immediate window (Figure 1.4) can be turned on by pressing **Ctrl+G**. VBA allows displacing immediate values, which are calculated during the numerical model. This is very helpful for debugging the program. One way is using the message box **Msgbox** (as demonstrated in Figure 1.2). The other way is using the **Debug. Print** tool. Unlike **Msgbox**, **Debug.Print** eliminates the need for any confirmation or acknowledgment every time. It is also used to show the values of variables by displaying a log of returned values in the immediate window.

The tool effectively evaluates a code to confirm that it is working properly and detect any bugs in the code. One of the common uses is to include the index and calculated values of each iteration within the loop to see if the iterations are going in the right direction.

1.3.4   BREAKPOINTS

The breakpoint specifies the line in your code where VBA should pause the execution of the macro operation when debugging the code. The shortcut key for adding a breakpoint is **F9**. Create a breakpoint, position the cursor on the specific line of your code where you want VBA to pause and press **F9**. Alternatively, locate the line of code where you want to add the breakpoint and click in the left grey margin beside the line. A dark red dot will appear on the margin beside the specified line to indicate that the breakpoint has been created. The specified line of code will also be highlighted with the same colour (Figure 1.4).

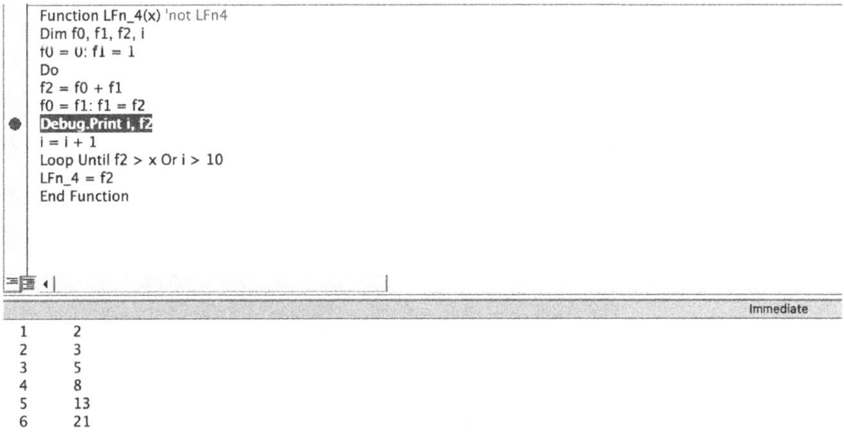

```
Function LFn_4(x) 'not LFn4
Dim f0, f1, f2, i
f0 = 0: f1 = 1
Do
f2 = f0 + f1
f0 = f1: f1 = f2
Debug.Print i, f2
i = i + 1
Loop Until f2 > x Or i > 10
LFn_4 = f2
End Function
```

```
                                                                  Immediate
1       2
2       3
3       5
4       8
5       13
6       21
```

FIGURE 1.4   Screenshot of the immediate window in VBE.

To remove the breakpoint, click on the dot again or press **F9** with the cursor positioned in the highlighted line of code. If there are many breakpoints, the user can remove them simultaneously by pressing **Ctrl+Shift+F9**.

Once the code runs to a breakpoint, VBA stops and highlights the breakpoint in yellow. Then the user can continue the code by pressing **F5**. Figure 1.4 shows the immediate values of the loop by inserting "**Debug.Print** i,f2" in function LFn_4 (Code 1.10).

1.4   USEFUL EXCEL TOOLS FOR ENGINEERING

Excel has two powerful calculation tools, Solver (Figure 1.5) and Goal Seek (Figure 1.6). These tools can be used to find the root of the equation and the maximum and minimum.

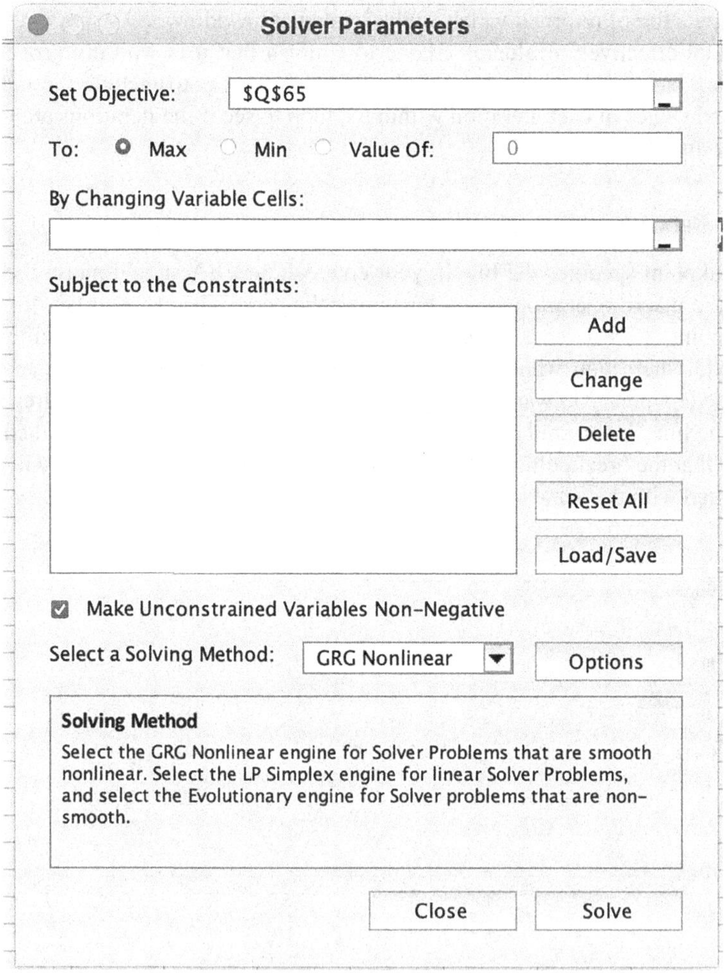

FIGURE 1.5   Screenshot of Solver.

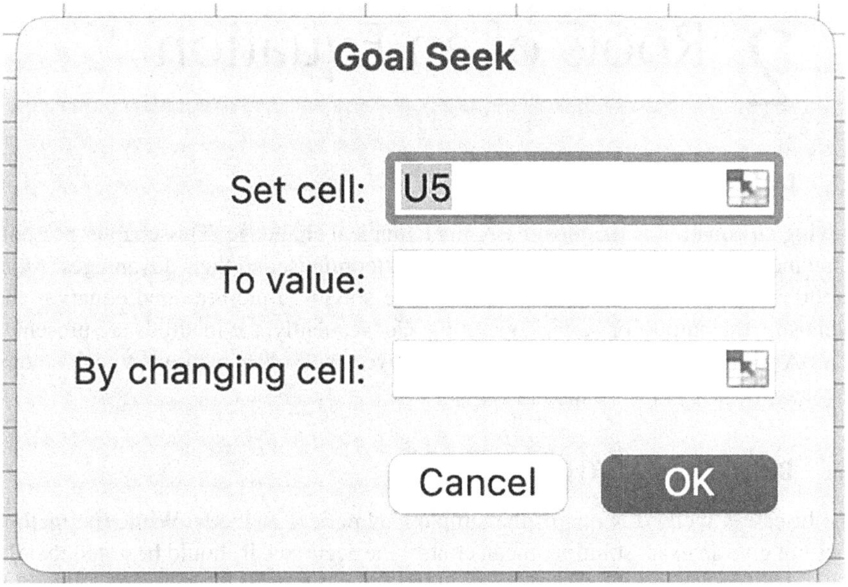

FIGURE 1.6   Screenshot of Goal Seek.

Solver has three algorithms. While the tools are practical in many cases, they do not show the underlying calculations. Thus, Solver and Goal Seek are not covered in this book.

2 Roots of an Equation

2.1 INTRODUCTION

Solving an equation is the most basic mathematical challenge. This chapter presents some methods to solve an equation, their shortcomings, and their advantages. More importantly, numerical methods often require solving a pre-arranged equation and then using the output for further operation. Consequently, the methods are presented in VBA forms, which can be routinely employed with other methods to solve complicated problems in subsequent chapters.

2.2 BISECTION METHOD

The bisection method is one of the simplest numerical methods. While the method does not give an exact solution, it can control the accuracy. It should be noted that the method does not solve for an exact root of the equation. Instead, the method identifies an interval that contains a root. The method controls the size of the interval. When the interval is diminished, the error becomes insignificant, and the root is the interval's midpoint. The method starts with an initial interval (x_1, x_2) with a change of sign. Subsequently, the value of the function at the midpoint, x_3, is calculated.

$$x_3 = \frac{x_1 + x_2}{2} \qquad \text{Eq. 2.1}$$

Depending on the value of $f(x_3)$, a new interval is selected, either (x_1, x_3) or (x_3, x_2). The halving procedure is repeated until the desired accuracy is met. The application will depend on the nature of the problem. In this chapter, the method is presented with flowcharts and codes. The methods can be combined with other functions to solve complicated problems. For example, the combined shooting/bisection method is used for heat transfer—the combined bisection/tear analysis for solving material balance with a recycling stream.

For coding, there are many variations of the method. Here a simple example is used to demonstrate these variations.

EXAMPLE 2.1.   SOLVING A NON-LINEAR EQUATION.

PROBLEM STATEMENT.

Find a positive root, between 0.5 and 2, of the following equation:

$$x^5 - 3x + 2\sin(x) = 0 \qquad \text{Eq. 2.2}$$

The required error of the solution is less than 0.001.

DOI: 10.1201/9781003364511-2

Solution

The halving procedure can be applied with three options in each iteration.

Code 2.1. Bisection method with a Do While loop.

```
Option Explicit
Function Bisection(a, b)
Dim x, Error, f1, f2, f3
Error = 0.001 'error
f1 = a ^ 5-3 * a + 2 * Sin(a): f2 = b ^ 5-3 * b + 2 *
Sin(b)
If f1 * f2 <= 0 Then 'The root is between a and b
Do While (b-a) > 2 * Error
x = (a + b) / 2 'A first estimate of the root by the
midpoint
f3 = x ^ 5-3 * x + 2 * Sin(x)
    If f3 = 0 Then
            GoTo SkipPoint
    ElseIf f1 * f3 < 0 Then 'The root is between a
and (a + b)/2
            b = (a + b) / 2 'Replace the upper estimate
    Else 'The root is between (a + b)/2 and b
            a = (a + b) / 2 'Replace the lower estimate
    End If
Loop
SkipPoint:
Bisection = x
Else
Bisection = "Wrong initial estimates" 'The root is
not between a and b
End If
End Function
```

There are some important notes on the VBA function in Code 2.1. First, the function is evaluated within one line to be typed directly. However, the function evaluation can be set as a separate VBA function and called in this function several times. Such an arrangement can help to reduce typos. Furthermore, the separate function can be called in an Excel spreadsheet directly and checked if the selected interval (a,b) contains a change-of-sign. Sometimes function evaluation requires multiple calculation steps and cannot be typed within one line. Hence, it is advised to set the equation as a separate VBA function, as in the following example.

The second noteworthy comment is about the first option in the **If** . . . **Elseif** structure, $f(x) = 0$. This option may help to reduce the number of iterations/calculations, but it rarely happens. If there are only two options, the code will be much simpler. Finally, the usage of SkipPoint: (or Exit Loop) is *not* recommended. Thus, the option $f(x) = 0$ should be removed in most cases.

2.2.1 BISECTION METHOD WITH CONTROL ERROR IN VARIABLE

If the required condition is an error in x, the number of iterations can be determined by

$$n = \frac{x_1 - x_2}{tolx}$$
Eq. 2.3

Hence, the **For . . . Next** structure can be used. The problem is solved, with the required error, *tolx*, being an argument, using the code to find a solution between (0.5, 2) with an error of 0.0001.

Code 2.2. A simple bisection using a For . . . Next loop.

```
Function BisectionX(Lo, Up, tolx)
Dim Mi, c1, c3, i, n
    n = Abs(Up-Lo)/tolx
'*****************************************
For i = 1 To n
    Mi = (Lo + Up) / 2
    c1 = f1(Lo): c3 = f1(Mi)
If c1 * c3 < 0 Then
    Up = Mi
Else
    Lo = Mi
End If
Next i
'*****************************************
BisectionX = Mi
End Function
Function f1(a)
f1 = a ^ 5-3 * a + 2 * Sin(a)
End Function
```

As the function is separated from the bisection function instead of using the direct formula as in Code 2.1. The previous format will be more useful for other cases. Notably, within each iteration, this VBA code requires two function evaluations, f1(Lo) and f1(Mi). The extra evaluations can be inefficient when the function evaluation is computationally intensive (see examples in the next chapter).

EXAMPLE 2.2.   DETERMINING THE FRICTION FACTOR FROM THE COLEBROOK EQUATION.

PROBLEM STATEMENT

The Colebrook equation is an implicit equation that combines experimental results of studies of turbulent flow in smooth and rough pipes (Shashi Menon 2015). For conduits that are flowing full of fluid at large Reynolds numbers, it is given as:

$$\frac{1}{\sqrt{f}} = -2log_{10}\left(\frac{\varepsilon / D_h}{3.7} + \frac{2.51}{Re\sqrt{f}}\right)$$ Eq. 2.4

where:
f is the Darcy friction factor, ε (m) is roughness height, D_h (m) is the inside diameter of the pipe and Re is the Reynolds number.
Part A. Calculate f with a required error less than 0.001 for the following conditions: $\varepsilon = 0.001$ m, $D_h = 0.05$ m and $Re = 4000$. The initial range is between 0.001 and 1.
Part B. Calculate f when the required difference between the left-hand and right-hand sides of the equation is less than 0.001.

Solution

For part A, the error is specified for the friction factor, f. Hence, the initial values can determine the number of iterations, and a For . . . Next loop structure (Figure 2.1) can be used. The While . . . Wend loop structure (Figure 2.2) can also be used, with the condition of $err > tolx$.

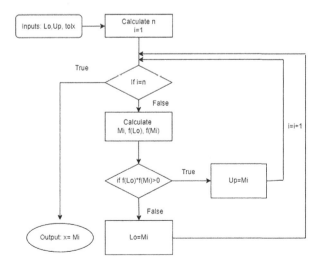

FIGURE 2.1   Flowchart for bisection code with a For . . . Next loop.

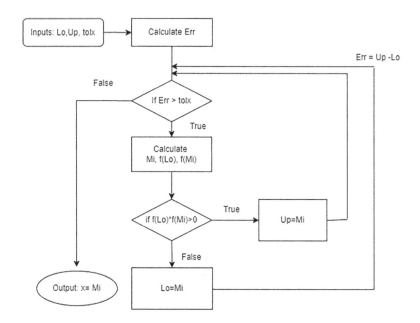

FIGURE 2.2   Flowchart for bisection code with a While . . . Wend loop.

Code 2.3. Bisection code with a For . . . Next loop and a single function evaluation.

```
Function BisectionX(Lower, Upper, e, D, Re)
Dim k, n, M, err, tol, c1, c2, c3
tol = 0.001
    n = (Upper-Lower)/tol
c1 = colebrook(Lower, e, D, Re)
For k = 1 To n
    M = (Lower + Upper) / 2: c3 = colebrook(M, e, D, Re)
If c1 * c3 < 0 Then
    Upper = M
Else
    Lower = M:c1=c3
End If
Next k
    BisectionX = M
End Function

Function colebrook(f, e, D, Re)
colebrook = (1 / f ^ 0.5) + 2 * Log((e/(D * 3.7)) +
(2.51 / (Re * (f ^ 0.5)))/Log(10))
End Function
```

The previous bisection code has removed the $f(x_3) = 0$ option. Furthermore, the calculation steps are rearranged so that each iteration requires only *one* function evaluation, $c_3 = f(x_3)$. It should be noted that the initial values have to be selected so that there is a change of sign. As a result, evaluation of c_2 is not required. However, a validation of sign change (such as in Code 2.1) might be needed in the general case. For parameter values, Code 2.3 returns $f = 0.010743043$. While the answer is sufficiently good, there is no guarantee that the difference between the left and right sides of Eq. 2.4 is smaller than 10^{-4}.

Code 2.4. Bisection code with limitation in function value, using the same objective function in Code 2.3.

```
Function BisectionF(Lower, Upper, e, D, Re) 'return f
Dim M, tol, c1, c2, c3, k
tol = 10 ^ -5
    M = (Lower + Upper) / 2
    c1 = colebrook(Lower, e, D, Re)
    c3 = colebrook(M, e, D, Re)
    k = 0
While Abs(c3) > tol And k < 100
If c1 * c3 < 0 Then
    Upper = M
Else
    Lower = M: c1=c3
End If
    M = (Lower + Upper) / 2: c3 = colebrook(M, e, D, Re)
k = k + 1
Wend
    BisectionF = M
End Function
```

Part B. To achieve the required accuracy, the error in function *colebrook* is used as the termination condition, as in Code 2.4. In this case, the number of iterations will depend on the specific tolerance of the function; thus, the number of iterations is unknown. Consequently, the For . . . Next loop is not suitable. Code 2.4 uses the While . . . Wend structure of Figure 2.2, with a small variation in testing conditions. The method can be applied to part A as well (by replacing Abs(c3) with Abs(Upper-Lower)).

2.2.2 BISECTION METHOD WITH SCANNING

In the previous examples, the bisection method needs an appropriate range before halving. The two limits are manually selected as initial guesses in the question. For a general problem, the selection of the two limits is not obvious. Consequently, it is advantageous to determine the limits identified via the scanning method. In a scanning procedure, only one starting point is needed. The scanning can be done with a constant or increasing step size. Hence scanning can identify the limits, which include a change of sign. Since the scanning procedure requires function evaluations, the last interval from the scanning step can be used as the starting region for the bisection method.

Code 2.5. Scanning and bisection code.

```
Function BisectionS(Lower, e, D, Re)
Dim step, Upper, M, tol, c1, c3, j, jmax
step = Lower / 2: tol = 10 ^ -7: jmax = 100
Upper = Lower + step
j = 0
Do Until c1 * c3 < 0 And j < jmax
step = step * 2
Lower = Upper: Upper = Lower + step
c1 = colebrook(Lower, e, D, Re): c3 = colebrook(Upper,
e, D, Re)
j = j + 1
Loop
If j = jmax Then
BisectionS = "There is no suitable interval for the
bisection method"
Else
M = (Lower + Upper) / 2: c3 = colebrook(M, e, D, Re)
While Abs(c3) > tol
If c1 * c3 < 0 Then
    Upper = M
    Else
    Lower = M: c1 = c3
End If
    M = (Lower + Upper) / 2: c3 = colebrook(M, e, D, Re)
Wend
    BisectionS = M
End If
End Function
```

The previous code can be applied with all the parameters in Example 2.1, with a starting point of 0.001. After the scanning part, the code will find a suitable interval between 0.0085 and 0.0165 after four scanning steps. In the case of multiple solutions, scanning can target the solution.

2.2.3 FAILURES OF THE BISECTION METHOD

There are special cases where the Bisection method fails. A classic example is the function:

$$f(x) = \frac{1}{x}$$

Eq. 2.5

The function has a change-of-sign interval between a positive and a negative point, yet it does not have a solution. Trying to solve this equation with Code 2.4 without a limit on the number of iterations will result in an endless loop.

Another kind of failure of bisection happens when the target function has a solution but no change-of-sign interval. An example function is:

$$f(x) = x^2 \qquad \text{Eq. 2.6}$$

This equation has a root $x = 0$, but both sides of the root are positive. There is an effective way to overcome these failures. In these cases, one can try to minimise the square of $f(x)$ within a specific range. A minimum of zero will indicate a root, whereas a non-zero minimum indicates no feasible root. To find a minimum of $f^2(x)$, one of the optimisation methods in Chapter 4 can be used.

2.3   NEWTON'S AND SECANT METHODS

The bisection method is straightforward but sometimes inefficient. Alternatively, there are other numerical methods to get the equation's root faster, requiring fewer iterations. Newton's method is one example. In this method, the next guessed value is determined from the function and first derivative of the function at the previous value. The iteration formula is:

$$x_{i+1} = x_i - \frac{f(x_i)}{f'(x_i)} \qquad \text{Eq. 2.7}$$

Unfortunately, the equation requires a derivative formula of the function f. In practice, this can be problematic. Some functions and correlations are not in the explicit form (such as the Colebrook equation in Eq. 2.4). Consequently, a finite-difference approximation of the derivative can be used:

$$f'(x_i) = \frac{f(x_i) - f(x_{i-1})}{x_i - x_{i-1}} \qquad \text{Eq. 2.8}$$

Hence, Eq. 2.7 can be reduced to:

$$x_{i+1} = x_i - f(x_i) \frac{(x_i - x_{i-1})}{f(x_i) - f(x_{i-1})} \qquad \text{Eq. 2.9}$$

The previous method is named the Secant method (Chapra and Canale 2010). While this method requires two values to calculate the following value (Figure 2.3), these two values should be close to each other (to avoid fluctuation and divergence). The termination condition can be based on the error in x or f. The solution will depend on the initial point and step size selection.

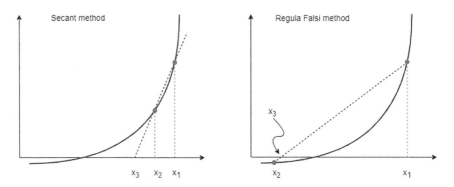

FIGURE 2.3   Two gradient-based methods, Secant and Regula Falsi.

A VBA function of the secant method is presented in Code 2.6., which is applied to the same Colebrook equation.

Code 2.6. Secant method.

```
Function secant(x1, x2, tol, e, D, Re)
Dim k, f1, f2, x3
f1 = colebrook(x1, e, D, Re): f2 = colebrook(x2, e, D, Re)
k = 0
While Abs(f2) > tol And k < 100
x3 = x2-f2 * (x2-x1)/(f2-f1)
x1 = x2: x2 = x3: f1 = f2: f2 = colebrook(x3, e, D, Re)
k = k + 1:
Wend
secant = x3
End Function
```

Using the code with $x_1 = 0.001$, $x_2 = 0.002$ and tol $= 10^{-7}$, one can get the same answer after eight iterations. It should be noted that the answer is extremely sensitive to the initial guesses. A bad selection, for example, 0.001 and 0.1, may lead to a negative value of x_3 and return #VALUE! (since the Colebrook function valuates $\log(x)$ and $x^{0.5}$).

2.4   THE REGULA FALSI METHOD

As seen previously, the Secant method can significantly reduce the number of iterations. However, it cannot verify whether it will lead to a solution. In that aspect, bisection is more reliable once the values at the two limits are validated. The Regula Falsi, or false position, is a compromise between the two approaches. Historically, this method was documented in cuneiform tablets from the Old Babylon period, 1800 BC (Chabert 1999).

The Regula Falsi method can be considered a mix between the bisection and secant methods. The method starts with a change-of-sign interval, as with the bisection method. Subsequently, the dividing point is selected by linear regression between the two points (Figure 2.3) instead of the mid-point. The iterative formula is:

$$x_3 = x_1 + (x_1 - x_2)\frac{f(x_1)}{f(x_1) - f(x_2)} \qquad \text{Eq. 2.10}$$

By comparing the signs of the $f(x_3)$, the next interval can be either (x_3, x_1) or (x_2, x_3). The linear regression formula can solve the friction factor problem. The method cannot be used with the For . . . Next loop because the number of iterations cannot be determined in advance.

Code 2.7. Regula Falsi method with a maximum limit on iterations.

```
Function PF(Lower, Upper, e, D, Re)
Dim M, tol, c1, c2, c3, k, output(1 To 2, 1 To 3)
tol = 10 ^ -7: k = 0
    c1 = colebrook(Lower, e, D, Re):
c2 = colebrook(Upper, e, D, Re)
    M = Lower-c1 * (Upper-Lower)/(c2-c1):
c3 = colebrook(M, e, D, Re)
While Abs(c3) > tol And k < 1000
If c1 * c3 < 0 Then
    Upper = M: c2 = c3
    Else
    Lower = M: c1 = c3
End If
    M = Lower-(Upper-Lower) * c1 / (c2 - c1):
c3 = colebrook(M, e, D, Re)
k = k + 1
Wend
PF = M
End Function
```

2.5 CONCLUDING REMARKS

There are many methods, with advantages and disadvantages, to numerically calculate a root of an equation. The selection will depend on the specific nature of the equation and the initial guesses. In general cases, either the bisection or false-position method is recommended, depending on the required error. If the control error is in the variable, use a bisection code (similar to Code 2.3). If the control error is in the function, the first choice should be the false-position method (Code 2.7).

2.6   FURTHER EXAMPLES FOR SELF-PRACTICE

EXAMPLE 2.3.   DETERMINE THE NUMBER OF ITERATIONS OF THE BISECTION CODE.

Modify the code in Codes 2.3 and 2.4 so that the number of iterations is shown in the intermediate windows.

Answer

Insert this line (after the loop).

```
Debug.Print k
```

For Code 2.3, k = 1000. For Code 2.4, k = 25.

EXAMPLE 2.4.   COMPARING EFFICIENCIES OF BISECTION AND REGULA FALSI METHODS.

Modify the code in Codes 2.4 and 2.7 to include the number of iterations in the output. Compare and discuss the results when the initial values are (0.001, 1) and (0.01, 1).

Answer

Change the function name to PF_2 and insert the following lines after the loop:

```
Dim output(1 To 2, 1 To 2)
output(1, 1) = "solution": output(2, 1) = M
output(1, 2) = "# iters": output(2, 2) = k
PF _ 2 = output
```

Also, do the same for the bisection method. The modified code will produce the number of iterations. Using the same initial range (0.001, 1) and the required error, 10^{-7}, the bisection method requires 30 iterations, while the Regula Falsi method requires 93 iterations. In this particular instance, the bisection method is three times faster. However, if the range is changed to (0.01, 1), the respective iterations are 27 and 7, which means the Regula Falsi method is about three times faster. The change in efficiency is due to the non-linear nature of the Colebrook function within this range. The Colebrook function for different values of f can also be calculated and plotted to demonstrate the correlation between f and the VBA function graphically.

3 Applications of Roots of an Equation

3.1 INTRODUCTION

The previous chapter demonstrates some numerical methods to solve a single explicit equation. This chapter presents some applications of the methods to engineering problems. In these problems, the nature of the correlation/data input is more complicated than an explicit equation. Therefore, the VBA code in the previous chapter is modified to adapt to these problems.

3.2 ADVANCED EXAMPLE: ADIABATIC FLAME TEMPERATURE

The adiabatic flame temperature is the highest temperature of combustion, in which all generated energy is used to heat the exhaust gases (Himmelblau and Riggs 2012). The determination of this temperature requires material and energy balances. First, the material balance is needed to determine the exhaust gas quantities. Second, the heat capacities are integrated to determine the change of enthalpy. Finally, the temperature is re-evaluated. Without a computer program, the temperature needs to be solved by a trial-and-error approach. In this section, VBA code is used to solve the problem. Furthermore, the advanced combination of code can determine the excess air at any given flame temperature.

EXAMPLE 3.1. DETERMINING THE ADIABATIC FLAME TEMPERATURE OF METHANE.

PROBLEM STATEMENT.

Consider the complete combustion of methane in an adiabatic reactor:

$$CH_4 + 2O_2 \rightarrow CO_2 + 2H_2O$$

Both fuel and air enter the reactor at 25°C. Heat capacity equations (Cp in kJ/gmol/°C and T in °C) for gases are determined from the following equation:

$$Cp = a + b \times T + c \times T^2 + d \times T^3 \qquad \text{Eq. 3.1}$$

The heat capacity constants and heat of formation are given in Table 3.1. Assume complete combustion and air containing 21% oxygen and 79% nitrogen.

Part A. Determine the adiabatic flame temperature of methane combustion with air; the air is supplied 50% excess.

Part B. Determine the required excess air so the adiabatic flame temperature is 1500°C.

DOI: 10.1201/9781003364511-3

TABLE 3.1

Heat Capacity Constants and Heat of Formation under Standard Conditions (Himmelblau and Riggs 2012)

	A	B	C	D	E	F
1		Heat of formation	Heat capacity constants			
2			a	b	c	d
3	CH_4	−49.963				
4	CO_2	−393.25	36.110	4.233E-02	−2.887E-05	7.464E-09
5	H_2O	−241.835	33.460	6.880E-03	7.604E-06	3.593E-09
6	O_2	0	29.100	1.158E-02	−6.076E-06	1.311E-09
7	N_2	0	29.000	2.199E-03	5.723E-06	−2.787E-09

Solution

Part A. For complete combustion, all fuel CH_4 reacts and forms CO_2. Using the basis of 1 gmol of CH_4, the amount of the gases is easily calculated. It should be noted that water is formed in vapour form. Following the material balances, the quantities of the gas output are:

$xCO_2 = 1$ gmol
$xH_2O = 2$ gmol
$xO_2 = 2x$ gmol
$xN_2 = 2 \times (79 / 21) \times (1 + x)$ gmol

In this system, x is the percentage of excess air. For part A, x equals 0.5. The heat of the reaction is calculated from the heat of formation:

$$\Delta H^{\circ}_{rxn} = \Delta \hat{H}^{\circ}_f\left(CO_2\right) + 2\Delta \hat{H}^{\circ}_f\left(H_2O\right) - \Delta \hat{H}^{\circ}_f\left(CH_4\right) = -826957 \frac{J}{mol} \quad \text{Eq. 3.2}$$

Consequently, the adiabatic flame temperature T_d is satisfied by the following equation:

$$\int_{25}^{T_d} \sum x_i Cp_i\left(T\right) dT = 826957 \quad \text{Eq. 3.3}$$

where $i = CO_2$, H_2O, O_2 and N_2.

The integration in Eq. 3.3 is applied to all species with the heat capacity given in Table 3.1. The values of the four coefficients are given in Table 3.1. Hence, three functions are developed (Code 3.1). The first VBA function calculates the sum of the integrations in Eq. 3.3 using temperature as an argument. The second VBA function, **ffT**, calculates the difference between the left and right sides of Eq. 3.3. The third VBA function calculates T_d so that **ffT** equals zero. Input A of the two VBA functions is the 4×4–cell range containing the heat capacity constants in Table 3.1. The code determines T_d at 1522.3°C when $x = 0.5$.

Code 3.1. Determining the adiabatic flame temperature of methane.

```
Option Base 1
Function enthalpyT(MC, T, xCO2, xH2O, xO2, xN2)
Dim i, ent(4)
For i = 1 To 4
ent(i) = MC(i, 1) * T + MC(i, 2) * T ^ 2 / 2 + MC(i,
3) * T ^ 3 / 3 + MC(i, 4) * T ^ 4 / 4
Next i
enthalpyT = ent(1) * xCO2 + ent(2) * xH2O + ent(3) *
xO2 + ent(4) * xN2
End Function


Function ffT(A, T, xCO2, xH20, xO2, xN2) 'note use
H20 not H2O within this section- still works fine
ffT = enthalpyT(A, T, xCO2, xH20, xO2, xN2)-
enthalpyT(A, 25, xCO2, xH20, xO2, xN2)-826957
End Function


Function BisectionT(Lo, Up, A, excess, tolx)
Dim Mi, c1, c2, c3, i, n, xCO2, xH20, xO2, xN2
     n = Abs(Up-Lo)/tolx
xCO2 = 1: xH20 = 2: xO2 = 2 * excess: xN2 = (79 / 21)
* 2 * (1 + excess)
c1 = ffT(A, Lo, xCO2, xH20, xO2, xN2)
For i = 1 To n
          Mi = (Up + Lo) / 2: c3 = ffT(A, Mi, xCO2,
xH20, xO2, xN2)
     If c1 * c3 < 0 Then
          Up = Mi:
     Else
          Lo = Mi: c1 = c3
End If
Next i
     BisectionT = Mi
End Function
```

Part B. The **BisectionT** function in Code 3.1 can determine T_d at a given excess, but the question requests the reverse correlation. Hence, another code can be combined with the previous VBA function. Code 3.2 (in combination with Code 3.1) produces $x = 0.531$. The combination takes a significant time if the limits are far apart.

It should be noted that this problem can be solved analytically by material balances. However, the calculation process is tedious with many constants. Writing two short VBA functions is more efficient (as we already have the VBA functions in Code 3.1). In this case, all VBA functions must be in the same Excel file. For example, the Excel spreadsheet needs to have heat capacity constants, as in Table 3.1. Using multiple VBA functions in this fashion is necessary for many complicated numerical problems. In this instance, the code used bisection on another bisection method. Since both functions require multiple iterations, an appropriate selection of initial values can significantly reduce the calculation time.

Code 3.2. Calculating air excess to control adiabatic flame temperature.

```
Function fffT(excess, Tmax, A)
fffT = BisectionT(1000, 2000, A, excess, 0.1)-Tmax
End Function
Function bisectionXT(Lo, Up, Tmax, A, tolg)
Dim Mi, c1, c2, c3, j, m
    m = Abs(Up-Lo)/tolg
    c1 = fffT(Lo, Tmax, A): c2 = fffT(Up, Tmax, A):
For j = 1 To m
    Mi = Lo + (Up-Lo) * c1 / (c1-c2)
    c3 = fffT(Mi, Tmax, A)
If c1 * c3 < 0 Then
        Up = Mi:
Else
        Lo = Mi: c1 = c3
End If
Next
    bisectionXT = Mi
End Function
```

3.3   ADVANCED EXAMPLE: TEAR ANALYSIS FOR MATERIAL BALANCES WITH A RECYCLE STREAM

In some cases, the objective function is not expressed in a formula. Instead, the evaluations are done via multiple steps. One of the examples is sequential modelling for process engineering. In this setup, the bisection method is applied to solve the system. For example, implicit correlation can be solved by the bisection method.

EXAMPLE 3.2.   TEAR ANALYSIS WITH A RECYCLE STREAM.

PROBLEM STATEMENT.

Ethyl chloride (CH_3CH_2Cl) can be manufactured by an exothermic reaction of hydrogen chloride (HCl) with ethylene ($CH_2{=}CH_2$):

$$C_2H_4 + HCl \rightarrow C_2H_5Cl$$

The process can be simplified in the following diagram, adapted from example 10.2–3 (Felder and Rousseau 2005). The fresh feed contains 50 kgmol/h of HCl and C_2H_4 each (equal molar flow). The fresh feed is mixed with the recycle stream before entering the reactor. The single-pass ethylene conversion is 90%. After the reactor, the stream is separated into Product (containing C_2H_5Cl only) and Stream S-4 (containing C_2H_4 and HCl). Stream S4 is split into Waste and Recycle streams. Streams S4, Waste and Recycle have the same composition.

Find the molar flow for the recycle stream if the waste stream has a flow rate of 8 kgmol/h.

Solution

First, a mathematical model is developed to describe all material balances of the process. Generally, each stream will have three chemicals (HCl, C_2H_4 and C_2H_5Cl). Since the flow rates of HCl and C_2H_4 are always equal on all streams, the unknown variables can be reduced to 1 and thus become a single-variable function. For other cases, the tear analysis may produce a multiple-dimensional system (see Example 8.2) and require a multiple-dimensional method.

The modelling of this process is based on a tear analysis. Tear analysis is performed by identifying recycle or loop streams in the process flow diagram and temporarily "tearing" them to create separate input and output streams. This simplifies the system into smaller sub-processes, allowing for more straightforward calculation and simulation. Once the steady-state conditions of these smaller sub-processes have been determined, the torn streams are reconnected.

For this particular problem, the tear stream can be either S1, S2, S4 or the recycle stream. For convenience, the recycle stream is selected. Hence, the flow rate of the recycle stream is assigned a random value, Rg. From Rg, the component flow rates in recycle stream are calculated. Subsequently, the materials of other streams are calculated from streams S1 to S2 to S4 (Figure 3.1). Consequently, the other side of the tear is calculated from S4, resulting in a value of $Rcal$. The next step of tear analysis is to adjust Rg so that $Rcal = Rg$. This can be done by a VBA function, as in Codes 3.3 and 3.4.

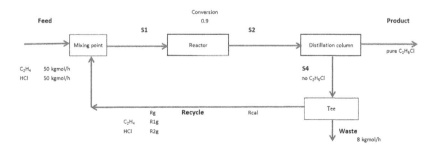

FIGURE 3.1   Acetylene production with a recycle stream.

Code 3.3. Calculate the difference between the guessed and calculated tear steam value (one variable).

```
Function diffR(Rg, feed, Conv, W)
Dim S1(3), S2(3), S4(3), Pro(3), Waste(3), R(3)
Dim Rcal
R(1) = Rg / 2: R(2) = Rg / 2 'calculate component of R
S1(1) = feed / 2 + R(1): S1(2) = feed / 2 + R(2)
S2(1) = S1(1) * (1-Conv): S2(2) = S1(2) * (1-Conv):
S2(3) = S1(1) * Conv 'we don't need S(3) for calc
S4(1) = S2(1): S4(2) = S2(2)
R(1) = S4(1)-W / 2: R(2) = S4(2)-W / 2:
Rcal = R(1) + R(2)
diffR = Rg-Rcal
End Function
```

The function in Code 3.3. calculates the difference between *Rg* and *Rcal*. In this code, the component flow rates of each stream are presented by an array of two elements. The flow rates are calculated from each unit operation's material balance.

Code 3.4. Solving tear analysis.

```
Function BisectionR(LowerR, UpperR, feed, Conv, W)
Dim x1, x2, x3, RelativeError, CalculatedError, f1,
f2, f3
RelativeError = 0.001 'Required relative error in %
CalculatedError = 1
x1 = LowerR: x2 = UpperR
    x3 = (x1 + x2) / 2 'A first estimate of the root
by the midpoint
If diffR(x1, feed, Conv, W) * diffR(x2, feed, Conv, W)
<= 0 Then 'The root is between L and U
Do While CalculatedError > RelativeError
    f1 = diffR(x1, feed, Conv, W): f3 = diffR(x3,
feed, Conv, W):
If f1 * f3 < 0 Then 'The root is between x1 x3
    x2 = x3
Else
    x1 = x3
End If
    x3 = (x1 + x2) / 2 'Estimate of the root by the
midpoint
    CalculatedError = Abs(x2-x1)
Loop
    BisectionR = x3
```

```
Else
     BisectionR = "Wrong initial estimates"
End If
End Function
```

By applying this code, the flow rate of the recycle stream is determined at $R = 2.22$ mol/h.

3.4 ADVANCED EXAMPLE: DETERMINING THE VAPOUR FRACTION FROM INTERNAL ENERGY AND SPECIFIC VOLUME

In some cases, the function cannot be evaluated by a formula, as in the previous examples. Instead, the data are in discrete points, often presented in a table. Consequently, the return values of variables can be evaluated from a data set by interpretation. VBA code can be developed for these cases to return the interpolated data. The principle of bisection and false-position methods is still applicable in combination with the linear interpretation.

EXAMPLE 3.3.   CALCULATING STEAM QUALITY AT A SPECIFIC ENERGY.

PROBLEM STATEMENT.

This problem is adapted from Example 9.16 (Himmelblau and Riggs 2012). In this case, the system contains a mixture of steam and saturated water liquid. The total internal energy and volume are specified:

$$\hat{U}_t = 322.9 \frac{Btu}{lb} \qquad \text{Eq. 3.4}$$

$$V_t = 6.25 \frac{ft^3}{lb} \qquad \text{Eq. 3.5}$$

The steam and vapour saturated values are given in Table 3.2.
Determine the composition (vapour fraction) and temperature of the mixture.

Solution

The unknowns of the process are vapour fraction x_v, total pressure P and temperature T. The degree of freedom of the system is zero. There is a unique solution, a specific combination of x_v, P and T, to satisfy the two conditions in Eqs. 3.4 and 3.5. On the other hand, the data are available in standard steam tables. In such tables, U and H are given for both phases as functions of T (or P). An example of a steam table is given in Table 3.2, which can be stored in an Excel spreadsheet.

TABLE 3.2
Steam Table for Saturated Vapour

Temperature (F)	Sat Pressure (lb/in²)	Specific Volume (ft³/lbm)		Internal energy (Btu/lbm)	
		Sat. Liquid	Sat. Vapour	Sat. Liquid	Sat. Vapour
32.018	0.08871	0.01602	3299.9	0	1021
35	0.09998	0.01602	2945.7	3.004	1019
40	0.12173	0.01602	2443.6	8.032	1015.6
45	0.14756	0.01602	2035.8	13.05	1012.2
50	0.17812	0.01602	1703.1	18.07	1008.9
55	0.21413	0.01603	1430.4	23.07	1005.5
60	0.25638	0.01604	1206.1	28.08	1002.1
65	0.30578	0.01604	1020.8	33.08	998.76
70	0.36334	0.01605	867.18	38.08	995.39
75	0.43016	0.01606	739.27	43.07	992.02
80	0.50745	0.01607	632.41	48.06	988.65
85	0.59659	0.01609	542.8	53.06	985.28
90	0.69904	0.0161	467.4	58.05	981.9
95	0.81643	0.01612	403.74	63.04	978.52
100	0.95052	0.01613	349.83	68.03	975.14
110	1.2767	0.01617	264.96	78.01	968.36
120	1.6951	0.0162	202.94	88	961.56
130	2.226	0.01625	157.09	97.99	954.73
140	2.8931	0.01629	122.81	107.98	947.87
150	3.7234	0.01634	96.929	117.98	940.98
160	4.7474	0.01639	77.185	127.98	934.05
170	5.9999	0.01645	61.982	138	927.08
180	7.5197	0.01651	50.172	148.02	920.06
190	9.3497	0.01657	40.92	158.05	912.99
200	11.538	0.01663	33.613	168.1	905.87
210	14.136	0.0167	27.798	178.15	898.68
212	14.709	0.01671	26.782	180.16	897.24
220	17.201	0.01677	23.136	188.22	891.43
230	20.795	0.01684	19.374	198.31	884.1
240	24.985	0.01692	16.316	208.41	876.7
250	29.844	0.017	13.816	218.54	869.21
260	35.447	0.01708	11.76	228.68	861.62
270	41.877	0.01717	10.059	238.85	853.94
280	49.222	0.01726	8.6439	249.04	846.16
290	57.573	0.01735	7.4607	259.26	838.27
300	67.028	0.01745	6.4663	269.51	830.25

Mathematically, the problem requires a combination of T and x_v ($0 < x_v < 1$) so that:

$$x_v \hat{U}_v + (1 - x_v) \hat{U}_l = \hat{U}_t \qquad \text{Eq. 3.6}$$

$$x_v V_v + (1 - x_v) V_l = V_t \qquad \text{Eq. 3.7}$$

The four values (U_l, U_v, V_v, V_l) are determined by linear interpolation from the table. The solution is based on the principle of the bisection method using two VBA functions. In the first function, the temperature is an input, and the corresponding total volume is the output. The function includes three steps:

1 For a given value of T, the corresponding U_v, U_l, V_v, V_l are obtained by interpolating data.
2 x_v is calculated by an interpolation formula of U_v, U_l to satisfy Eq. 3.6, $U = 322.9$ Btu/lbm. Note that if $U_v < U_t$ or $U_l > U_t$, there is no physically feasible value of x_v. However, the formula can assign $x_v = 0$ and $x_v = 1$ under those conditions.
3 The total volume V_t is calculated by Eq. 3.7 from x_v and V_v, V_l.

A bisection (or false-position method) is then applied to find T so that the previous function, function T_x in Code 3.5, equals V_t. Finally, the determined T_x is combined with another interpolation to find P.

Code 3.5. Calculating steam temperature and vapour fraction. The discrete data (steam table) are stored in cells A2:F100 of the active spreadsheet.

```
Function Tx(t, U)
Dim Ul, Uv, i, fraction, volume, output(1, 1)
output(0, 0) = "total V": output(0, 1) = "Vapour fraction"
For i = 3 To 200
If t <= Cells(i + 1, 1) And t >= Cells(i, 1) Then
Ul = Cells(i, 5) + (t-Cells(i, 1))/(Cells(i + 1, 1)-
Cells(i, 1)) * (Cells(i + 1, 5)-Cells(i, 5))
Uv = Cells(i, 6) + (t-Cells(i, 1))/(Cells(i + 1, 1)-
Cells(i, 1)) * (Cells(i + 1, 6)-Cells(i, 6))
Select Case U
Case Is < Ul
    fraction = 0
Case Is > Uv
    fraction = 1
Case Else
    fraction = (U-Ul)/(Uv-Ul)
End Select
output(1, 0) = Cells(i, 4) * fraction + Cells(i, 3) *
(1-fraction)
output(1, 1) = fraction:
```

```
Tx = output
Exit Function
End If
Next i
Tx = 2000
End Function


Function BisectionTx(Lo, Up, tolt, U, V)
Dim Mi, c1, c3, i, n, ent1, ent3, output(1, 1)
    n = Abs(Up-Lo)/tolt
ent1 = Tx(Lo, U): c1 = ent1(1, 0)-V
For i = 1 To n
    Mi = (Lo + Up) / 2: ent3 = Tx(Mi, U): c3 = ent3(1, 0)-V
If c1 * c3 < 0 Then
            Up = Mi
Else
            Lo = Mi: c1 = c3
End If
Next i
output(0, 0) = "Temperature": output(0, 1) = "Vapour
fraction"
output(1, 0) = Mi: output(1, 1) = ent3(1, 1)
BisectionTx = output
End Function
```

In this case, the vapour and enthalpy are determined via interpolation (instead of the explicit equations from P or T). Using this code, the solution is $T = 210°C$ and $x = 0.20$. It should be noted that the function **Tx** returns a 2×2 matrix. The following bisection code only requires an element of this output matrix. Since the types of two variables, *ent1* and *ent3*, are not specified, the VBA function assigns the whole 2×2 matrix to these variables. Subsequently, element (1,0) is used for the bisection procedure. The method may not work for an array of other dimensions (see Example 3.6).

3.5 FURTHER EXAMPLES FOR SELF-PRACTICE

EXAMPLE 3.4. COMBINATION OF BISECTION AND FALSE-POSITION METHODS.

Change Code 3.2 to the false-position method. Note the number of iterations.

Answer

The false-position method needs only one iteration. The reason is that the T_d-versus-x correlation is linear.

EXAMPLE 3.5.   AN ALTERNATIVE SOLUTION TO THE STEAM PROBLEM IN EXAMPLE 3.3.

In step 2, x_v can be found by interpolating the specific volume (instead of specific energy). What is the final answer?

Answer

The answers are $Td = 206.7°C$ and $x_v = 0.21$. These are more accurate than the previous one. The reason is that the linearity of U-versus-T is better than V-versus-T. The solution can be improved using the curve fitting for four values in Table 3.2.

EXAMPLE 3.6.   USING A ONE-DIMENSIONAL ARRAY AS THE OUTPUT.

Redo Code 3.5 using the output of function Tx as a 1-dimensional array and apply it to the bisection code. What would happen?

Answer and discussion

Replace output(1,1) with output(1).

```
output(0) = Cells(i, 4) * fraction + Cells(i, 3) *
(1-fraction)
output(1) = fraction:
```

For the second function, use the following:

```
ent1 = Tx(Lo, U): c1 = ent1(0) - V
```

The first function, Tx, still returns a correct answer (2×1 matrix) in an Excel spreadsheet. The second function returns #VALUE. Maybe other Excel versions can work with a 1-dimensional array! One way to overcome the error is using output (0,3).

EXAMPLE 3.7.   DETERMINE THE COMPOSITION OF A BINARY MIXTURE FROM THE REFRACTIVE INDEX.

The refractive index of the water-ethanol mixture is given in Table 3.3 (Herráez and Belda 2006). Create VBA codes to find mole fraction, x, as a function of the ethanol fraction, n_D. Apply the code for finding x for $n_D = 1.3532$.

TABLE 3.3

The Refractive Index of the Water-Ethanol Mixture

Ethanol Mole Fraction	Refractive Index, n_D
0	1.3326
0.101	1.3457
0.2001	1.3535
0.2987	1.3574
0.3332	1.3582
0.4087	1.3595
0.5051	1.3604
0.603	1.3614
0.6664	1.3618
0.7032	1.3617
0.7956	1.3614
0.9075	1.3609
1	1.3593

Hint

Modify the first function in Code 3.5.

4 One-Dimensional Optimisation

4.1 INTRODUCTION

Finding the optimum, either maximum or minimum, is a common problem in engineering. The objective function can be cost, material, energy or time. In comparison, the input can be any physical variable. This chapter will deal with numerical optimisation. Analytically, the optimum of an equation can be found to differentiate the equation and solve for the root, $f'(x) = 0$. Various methods in the previous chapter can find the root of the equation. However, such an approach has two problems. First, one must clarify whether the optimum is a minimum or a maximum. Second, the differentiation might be challenging or impossible (for example, implicit equations). The numerical methods in this chapter will overcome those challenges.

4.2 REGION ELIMINATION

Similar to the bisection/Regula Falsi methods, the region elimination approach can be applied to optimisation problems. The method starts with an initial region, which contains the optimum. Consequently, two internal points are selected, and the values of the target function at these two points are compared. Depending on the relative values and nature of the optimisation, a section of the original region is eliminated. The iteration is repeated until the required tolerance is met. The comparison is simple.

Let a and b be initial ends ($a < b$). Two values, x_1 and x_2, are selected between a and b ($x_1 < x_2$). The function is evaluated at these two values for comparison. For minimisation problems, the three elimination options are:

- If $f(x_1) < f(x_2)$: the region (x_2, b) is eliminated; x_2 becomes the new b for the next iteration.
- If $f(x_1) > f(x_2)$: the region (a, x_1) is eliminated; x_1 becomes the new a for the next iteration.
- If $f(x_1) = f(x_2)$: two regions (a, x_1) and (x_2, b) are eliminated; x_1 and x_2 become new a and b for the next iteration, respectively.

For maximisation, the opposite of conditions (1) and (2) is applied. Option 3 is sometimes omitted (Figure 4.1). Similar to the bisection method, there are two important shortcomings of the elimination method. First, it does not guarantee that an optimum exists. Second, it does not guarantee global optimum within the region

DOI: 10.1201/9781003364511-4

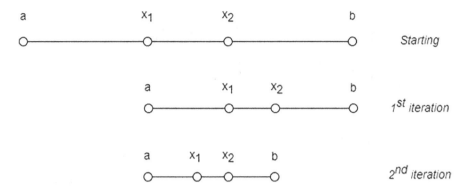

FIGURE 4.1   Golden section search.

(a,b). Generally, a graphic demonstration of the function can be used to verify the number of optimums between (a,b).

In the literature, there are three ways to select the two interior points: (i) binary search, (ii) Fibonacci search and (iii) golden section search. In the first search, the two interior points are selected as close as possible to the midpoint of the region. In this case, each iteration eliminates nearly half of the original region. While the binary search is the most efficient way to divide the region, it requires two function evaluations at each iteration. For some problems, the function evaluation is a time-consuming step. Consequently, reducing the number of function evaluations while slightly increasing the number of iterations is preferable.

The other two methods, Fibonacci and golden section, require only one function evaluation per iteration (except the first iteration). Even though these two methods require more iterations than the binary search, they might reduce the computing time. In these instances, the two dividing points are selected so that one of them would be the next dividing point of the subsequent iteration. The selection is compensated by reducing the elimination efficiency and increasing the number of iterations. Fibonacci has the lowest number of iterations but requires the series of Fibonacci numbers. The golden section method relies on the "golden ratio", which has been studied since Euclid (Chapra and Canale 2010). While Excel/VBA does not have an in-built function for the golden ratio, it can be calculated as $(1+5^{0.5})/2$. After each iteration, the region is reduced by 0.382.

With a predetermined error (the width of the final region), the number of iterations of the three methods can be determined. Hence the For Next loop can be used. Among the three elimination methods, however, the golden section method is the only one that does not require a predetermined error to calculate the two interior points (Chapra and Canale 2010). In this section, VBA code for the golden section is shown. Furthermore, the combination of the golden section method will be demonstrated in Chapter 10.

EXAMPLE 4.1.   MINIMISATION BY GOLDEN SECTION SEARCH.

PROBLEM STATEMENT.

The operating cost of a pump and a motor is given:

$$f(x) = 400 + 1.2x + \frac{25}{x} \qquad \text{Eq. 4.1}$$

where x is the motor power, expressed in horsepower (h.p.).

The manufacturer can provide a size from 1 to 8 h.p. with an increment of 0.1 h.p. Find the optimum pump size to minimise the cost of operating a pump and motor.

Solution

In this problem, we need to find a minimum of Eq. 4.1 between 1 and 8. Since the size must be a multiple of 0.1, the error in x should be 0.1. To get the nearest number with one digit after the decimal point, the in-built function **Round** is used.

Code 4.1. Golden section search.

```
Function GoldenSection _ R(Dlower, Dupper)
Dim D1, D2, f1, f2, err, i, R
err = Abs(Dupper-Dlower)
R = (5 ^ 0.5-1) / 2
i = 0
D2 = Dlower + R * (Dupper-Dlower): D1 = Dupper-R *
(Dupper-Dlower)
f1 = ff(D1): f2 = ff(D2)
While err > 0.1 And i < 100
i = i + 1
    If f1 > f2 Then
    Dlower = D1: f1 = f2: D1 = D2
    D2 = Dlower + R * (Dupper-Dlower): f2 = ff(D2)
    Else
    Dupper = D2: f2 = f1: D2 = D1
    D1 = Dupper-R * (Dupper-Dlower): f1 = ff(D1)
    End If
err = Abs(Dupper-Dlower)
Wend
```

```
GoldenSection _ R = WorksheetFunction.Round((D1 + D2)
/ 2, 1)
End Function
Function ff(x)
ff = 400 + 1.2 * x + 25 / x
End Function
```

The code returns an answer of 4.6 after 9 iterations. Without using **Round**, the code will return an answer of 4.5569.

4.3   QUADRATIC INTERPOLATION

The previous methods only compare the values of internal points without considering the magnitude of these values. The elimination can be very inefficient. The quadratic interpolation can take the relative slope of the values into the optimisation. The method starts with three points and assumes that the curve follows a quadratic function. Subsequently, the method calculates the bottom of the parabola for a minimum. The method starts with three initial values. The next value is determined for each iteration by fitting the quadratic equation. Let us name the three initial points as x_1, x_2, x_3 $(x_1 < x_2 < x_3)$, and the three corresponding evaluations are f_1, f_2, f_3 (for minimisation, f_2 is smaller than both f_1 and f_3). The peak of the parabola is:

$$x_4 = \frac{1}{2} \frac{f_1\left(x_2^2 - x_3^2\right) + f_2\left(x_3^2 - x_1^2\right) + f_3\left(x_1^2 - x_2^2\right)}{f_1\left(x_2 - x_3\right) + f_2\left(x_3 - x_1\right) + f_3\left(x_1 - x_2\right)} \qquad \text{Eq. 4.2}$$

The new point is combined with two of the previous points. Then, either x_1 or x_3 will be replaced. The selection is based on the position of x_4. Once the three new values are updated, the iteration can be repeated.

EXAMPLE 4.2.   MINIMISATION BY QUADRATIC INTERPOLATION.

Problem statement. Repeat Example 4.1 by quadratic interpolation.

Solution

Code 4.2. Quadratic interpolation.

```
Function quadratic(x1, x2, x3, tol) 'note x2 is
between x1-x3
Dim f1, f2, f3, err, i, x4
err = Abs(x3-x1)
```

```
While err > (tol / 2) And i < 100
i = i + 1: f1 = ff(x1): f2 = ff(x2): f3 = ff(x3)
x4 = 0.5 * (f1 * (x2 ^ 2-x3 ^ 2) + f2 * (x3 ^ 2-x1 ^ 2)
+ f3 * (x1 ^ 2-x2 ^ 2))/(f1 * (x2-x3) + f2 * (x3-x1)
+ f3 * (x1-x2))
Select Case x4
    Case Is >= x2
    x1 = x2: x2 = x4
    Case Is < x2
    x3 = x2: x2 = x4
    Case Else
    quadratic = "No solution"
    End Select
err = Abs(x1-x3)
Wend
quadratic = Round((x1 + x3) / 2, 1)
End Function
```

With $x_1 = 1$, $x_2 = 2$ and $x_3 = 8$, the code returns the same answer after six iterations.

4.4   OPTIMISATION WITH SCANNING

Similar to the bisection method, region elimination needs two good guesses. Hence, one can combine the scanning method (such as Code 2.5) with one of the previous methods. It is also essential to have an appropriate initial range. For example, to find a minimisation by quadratic method, the three initial starting points need to be in the proper order: $f(x_2)$ is smaller than $f(x_1)$ and $f(x_3)$. The three points can be routinely achieved by a scanning method. A simple scanning macro can be developed and combined with the previous quadratic interpolation (Code 4.3).

Code 4.3. Quadratic interpolation with scanning (combined with Code 4.2).

```
Function scanning(x1, delx, tol)
Dim i, x2, x3, f1, f2, f3
x2 = x1 + delx: f1 = ff(x1): f2 = ff(x2): i = 0
While f1 > f2 And i < 1000
x1 = x2: f1 = f2: delx=delx*2: x2 = x1 + delx:
f2 = ff(x2)
i = i + 1
Wend
If i = 1000 Then
    scanning = "no minimum"
```

```
Else
     scanning = quadratic(x1 - delx/2, x1, x2, tol)
End If
End Function
```

The code starts with a small initial value and requires no assumption on the upper limit. The code also doubles the scanning step after each iteration. The code stops the scanning procedure when a region containing a minimum, a combination of three points satisfying the initial conditions of the quadratic interpolation, is found. With $x_1 = 0.1$ and *delx* = 0.2, the code stops scanning after five iterations and starts interpolating with three values (3.1, 6.3, 12.7).

In the previous example, the VBA function scanning calls the quadratic interpolation at the end. A similar approach can be applied to the golden section search and other region elimination methods. While these methods only require two initial points, it is essential to know that at least there is an optimum between those two points (that is, an interior point whose function is smaller than the function values at two ends).

4.5   ADVANCED EXAMPLE: A COMBINATION OF BINARY SEARCH AND BISECTION METHOD

Optimisation can be done with a complicated function, such as an implicit function. In such instances, the solving equation may require another numerical program, such as a bisection or secant method.

EXAMPLE 4.3.   MINIMISATION OF FRICTION LOSS FACTOR.

PROBLEM STATEMENT.

Part A. For fluid with a kinetic viscosity of 0.00123 m²/s and flow of 2 m³/s, calculate the friction loss, *f*, as a function of pipe diameter between 0.04 and 0.6 m. The roughness is assumed at 0.001 m.
Plot the results.
Part B. Find the minimum value of *f* and the corresponding diameter.

Solution

Part A. Here, the Colebrook function of Example 2.1 is employed again. Since the equation is in implicit form, a bisection method is needed. However, we want to see the correlation between *f* and *D*. As *D* increases, velocity and Reynolds number decrease. Consequently, the value of *f* changes. A numerical method is needed to find the optimum value of *f*.

First, fundamental equations of fluid mechanics are needed. For a fluid flow in a circular pipe (diameter D), the Reynolds number is defined as:

$$Re = \frac{VD}{v}$$

Eq. 4.3

where V is the velocity (m/s), and v is the kinetic viscosity (m²/s).

The velocity is a function of the volumetric flowrate (Q) and cross-section area:

$$V = \frac{4Q}{\pi D^2}$$

Eq. 4.4

The previous equations can be combined with the Colebrook equation to determine the friction factor. Bisection code similar to the previous one is used for Part A. However, the inputs include D and Q (instead of D and Re, as in Code 2.5).

Code 4.4. Bisection method to find friction loss as a function of the pipe diameter.

```
Function Bisection2(Lower As Double, Upper As Double,
e, D, Q)
Dim M, err, tol, c1, c2, c3, V, Re 'V and Re are
variables
V = Q * 4 / (D ^ 2 * 3.1416)
Re = D * V / 0.00123
tol = 10 ^ -15
    c1 = colebrook(Lower, e, D, Re)
    M = (Lower + Upper) / 2: c3 = colebrook(M, e, D, Re)
err = abs(Upper - Lower)
While err > tol
If c1 * c3 = 0 Then
    Bisection2 = M
    Exit Function
Else
    If c1 * c3 < 0 Then
    Upper = M
    Else
    Lower - M: c1 = c3
    End If
End If
    M = (Lower + Upper) / 2: c3 = colebrook(M, e, D, Re)
err = Upper-Lower
Wend
    Bisection2 = M
End Function
```

The previous code can be applied to values of D from 0.04 to 0.6 and is plotted in Figure 4.2. The curve indicates that the correlation has a minimum of around $D = 0.2$ m.

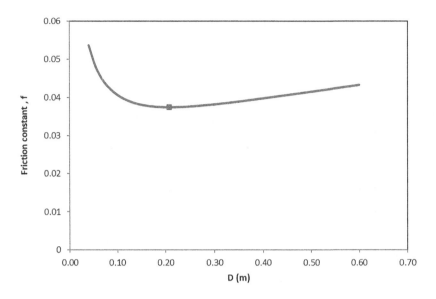

FIGURE 4.2   Friction loss as a function of velocity, as shown in Example 4.2.

Part B. To get the minimum in Figure 4.2, a binary search can be used.

Code 4.5. Combined binary search and bisection methods (the Bisection2 function is in Code 4.4).

```
Option Base 0
Function BinaryFunction(Dlower, Dupper, Q, e)
Dim V, Re, tol, D1, D2, B1, B2, i, err
tol = 10 ^ -7: err = Abs(Dupper-Dlower)
i = 0
While err > tol And i < 50
D1 = (Dupper-Dlower) / 2-tol + Dlower: D2 = (Dupper-
Dlower) / 2 + tol + Dlower
B1 = Bisection2(0.0001, 1, e, D1, Q):
B2 = Bisection2(0.0001, 1, e, D2, Q)
i = i + 1
If B1 = B2 Then
Dlower = D1: Dupper = D2
Else
     If B1 > B2 Then
     Dlower = D1
     Else
     Dupper = D2
     End If
End If
```

```
err = Abs(Dupper-Dlower)
Wend
BinaryFunction = (D1 + D2) / 2
End Function
```

Using the previous code, $D^* = 0.207$. To get the minimum of f, one can call the function **Bisection2** and get $f_{min} = 0.007059$. The code can run for a substantially long time if the initial region is large. One of the disadvantages of the previous bisection code is that two initial limits are fixed. These two limits are unnecessarily wide apart to ensure a solution exists in all values of D. The large initial region will require many iterations during the bisection method. Applying the scanning method with double step size might significantly reduce the computing time in this case.

4.6 FURTHER EXAMPLES FOR SELF-PRACTICE

EXAMPLE 4.4. COMPARISON BETWEEN BINARY SEARCH AND GOLDEN SECTION SEARCH.

Calculate the number of iterations between different search methods.

Answer

Insert the index into the code so the program gives the number of iterations.
For example, for Code 4.1, change the last line to

```
Dim output(1,2)
output(1,1) = Round((D1 + D2) / 2, 1)
output(1,2) = i
GoldenSection =output
```

EXAMPLE 4.5. FIBONACCI SEARCH.

Write VBA code for binary search and Fibonacci code.

Hints

First, the Fibonacci numbers are generated up to the required limit: $(b-a)/\varepsilon$, where ε is the tolerance. In Example 4.1, the limit is 80. So the required Fibonacci numbers are 1, 1, 2, 3, 5, 8, 13, 21, 34, 55, 89.

For each iteration, the selection of x_1 and x_2 is based on the two neighbouring Fibonacci numbers: $x_1 = b - F_i/F_{i+1}(b-a)$ and $x_2 = a + F_i/F_{i+1}(b-a)$. For example, the first iteration is based on F_{10} (55) and F_{11} (89).

EXAMPLE 4.6.   CHANGE THE OUTPUT OF THE BINARY SEARCH

Change Code 4.5 so that the binary code returns the value of the corresponding D (instead of f).

Hints

Modify the last line of the code "BinaryFunction $= (B1 + B2)/2$". Alternatively, the code can give f_{min} and $D*$ by employing an output matrix, as in Example 4.4.

EXAMPLE 4.7.   CHANGE THE REQUIRED INCREMENT IN OUTPUT.

Repeat Example 4.1, with a different requirement: the manufacturer can provide a size from 1 to 8 h.p. with an increment of 0.25 h.p.

Solution

Excel has a rounding function, MROUND, which allows rounding a number to a specified multiple. For example, using the following code with $tol = 0.25$ will result in $D = 4.5$.

```
Function GoldenSection(Dlower, Dupper, tol)

Dim D1, D2, f1, f2, err, i, R
err = Abs(Dupper-Dlower)
R = (5 ^ 0.5-1) / 2
i = 0

D2 = Dlower + R * (Dupper-Dlower): D1 = Dupper-R *
(Dupper-Dlower)
f1 = ff(D1): f2 = ff(D2)

While err > tol And i < 100
i = i + 1
    If f1 > f2 Then
```

```
    Dlower = D1: f1 = f2: D1 = D2
    D2 = Dlower + R * (Dupper-Dlower): f2 = ff(D2)
    Else
    Dupper = D2: f2 = f1: D2 = D1
    D1 = Dupper-R * (Dupper-Dlower): f1 = ff(D1)
    End If
err = Abs(Dupper-Dlower)
Wend
GoldenSection = WorksheetFunction.Mround((D1 + D2) /
2, tol)
End Function
```

5 Linear Programming

5.1 INTRODUCTION

While the previous methods in Chapter 4 deal with one-dimensional optimisation, on the other hand, many practical problems require multiple-dimensional optimisation. In general cases, these multiple-dimensional problems are challenging to solve. However, there is a class of optimisation in which all equations are linear. This type of optimisation can be solved by "linear programming" algorithms. The most popular one is the simplex algorithm.

The simplex algorithm is widely used for solving linear programming problems and is still considered one of the most efficient methods for solving linear programming problems. A linear programming problem is an optimisation problem where we seek to maximise or minimise a linear objective function subject to a set of linear constraints. The simplex algorithm works by iteratively moving from one basic feasible solution to another in a way that maximises or minimises the objective function. One of the strengths of the simplex algorithm is its ability to handle large-scale linear programming problems.

5.2 SIMPLEX ALGORITHM

The simplex algorithm is better demonstrated in the table form. All equations, including the objective function, are converted into a table form. The problem can be summarised by using the following example:

EXAMPLE 5.1.   MAXIMISE A LINEAR FUNCTION.

PROBLEM STATEMENT.

Maximise the following function:

$$z = 20x_1 + 16x_1 + 12x_3 + 10 \qquad \text{Eq. 5.1}$$

Subject to:

$$x_1 < 4$$
$$2x_1 + x_2 + x_3 < 10$$
$$2x_1 + 2x_2 + x_3 < 16$$

All variables are non-negative.

 DOI: 10.1201/9781003364511-5

Solution

The first step is to convert the inequalities to equalities and then present the table form. The inequality constraints are converted to equations by adding slack variables.

$$x_1 + x_4 = 4 \qquad \text{Eq. 5.2}$$

$$2x_1 + x_2 + x_3 + x_5 = 10 \qquad \text{Eq. 5.3}$$

$$2x_1 + 2x_2 + x_3 + x_6 = 16 \qquad \text{Eq. 5.4}$$

The objective function is rewritten:

$$-z + 20x_1 + 16x_1 + 12x_3 = -10 \qquad \text{Eq. 5.5}$$

Consequently, the objective and equations can be presented in the matrix form. The simplex algorithm can be applied by improving the feasible solution. The problem is presented in the matrix form, Table 5.1, and entered into an Excel spreadsheet.

To simplify the explanation, the variables can be classified as either basic or non-basic variables. In Table 5.1, the three variables without coefficients (x_4, x_5, x_6) are *basic* variables. The other variables (x_1, x_2, x_3) are *non-basic* variables. The simplex method employs pivoting steps in a repeating manner. The purpose of each pivoting step is to exchange one basic variable and one non-basic variable (Nawrocki 2013). The pivoting includes three elementary steps:

1. Determine the non-basic variable to be pivoted. Find the largest positive C (in row 1) and the corresponding column index (AcC). If all values in row 1 are negative, stop the iteration.
2. Determine the basic variable to be exchanged. Within the column AcC, from A(AcC,2) to A(AcC,m), find row AcR so that B(AcR)/A(AcC, AcR) has the smallest positive value. If all A(i, AcC) are non-positive, stop (that problem is unbounded).
3. Pivot on the basic variable, A(r,c): after this step, x(AcC) becomes the basic variable, and one of the basic variables becomes a non-basic variable.

TABLE 5.1
The Input of the Simplex Problem

-z	x_1	x_2	x_3	x_4	x_5	x_6	B
1	20	16	12				-10
	1	0	0	1	0	0	4
	2	1	1		1		10
	2	2	1			1	16

The pivoting is repeated until all c are negative. It should be noted that cycling can happen when steps 2 and 3 have several equal choices. To prevent cycling, one can use Bland's anticycling rule:

- Select largest a(1,c) with smallest c (first value)
- Select smallest a(1,c) with smallest r (first value).

The procedure is implemented in the following code. The input is a table (containing all elements in Table 5.1 except the headings), and the output is a table of the same size. The code is divided into three parts for the readers to follow. The first one is validating the dimension and properties of the inputs.

Code 5.1. Linear programming, section 1. Declaring dimensions and checking column B for negative numbers.

```
Function LinearP(A)
Dim output(), checkB(), checkC(), checkR(), AcC, AcR,
MinC, MinB, i, j, temp, k, P1, P2
Dim M(), nRow, nCol
Dim RowDim As Integer
Dim ColDim As Integer
nRow = A.Rows.Count: nCol = A.Columns.Count
ReDim checkB(2 To nRow): ReDim checkR(1 To nRow) 'to
find the r (leaving variable)
ReDim checkC(1 To nCol) 'to find the column (enter-
ing variable)
ReDim M(1 To nRow, 1 To nCol) 'M is output matrix,
same size as A
For i = 1 To nRow
    For j = 1 To nCol
          M(i, j) = A(i, j)
    Next j
Next i
'check B, from 2 to n for a negative number
For i = 2 To nRow
    checkB(i) = A(i, nCol)
Next i
MinB = WorksheetFunction.Min(checkB())
If MinB < 0 Then
    LinearP = "negative B"
    Exit Function
End If
```

After checking the matrix, the code needs to check if there is a positive coefficient in the first row (Code 5.2). If there is none, the current solution cannot be improved.

Code 5.2. Linear programming, section 2: determining AcC (entering non-basic variable) and AcR (row to be pivoted).

```
For j = 2 To (nCol-1)
If M(1, j) > 0 Then
    checkC(j) = M(1, j)
End If
Next j 'if there are equal M(1,j), the last (largest
j) is selected
MinC = WorksheetFunction.Max(checkC()) 'Debug.Print MinC
If MinC > 0 Then
k = 0
Do While (MinC > 0) And (k < 10)
temp = 0
    For j = 2 To (nCol-1)
            If MinC = M(1, j) Then
                    AcC = j
            End If
    Next j:
ReDim checkR(1 To nRow) 'Erase checkr() set to zero,
ReDim sets to empty
    For i = 2 To nRow
            If (M(i, AcC) > 0) Then
                    checkR(i) = M(i, nCol)/M(i, AcC)
            End If
    Next i
temp = WorksheetFunction.Min(checkR()):
    For i = 2 To nRow
            If temp = checkR(i) Then
                    AcR = i
            End If
    Next i
```

In the next step, a basic and non-basic variable will be determined to be exchanged with each other (Code 5.3).

Code 5.3. Linear programming, section 3. Pivoting.

```
'cell (AcR, AcC) is the target- normalise ActiveRow
so that M(AcR,AcC) equals to 1
P1 = M(AcR, AcC): 'Debug.Print AcR, ratio1, M(nrow,
ncol)
For j = 2 To nCol
```

```
M(AcR, j) = M(AcR, j)/P1
Next j: 'Debug.Print AcR, AcC, M(AcR, AcC), M(AcR,
ncol)

For i = 1 To nRow 'pivoting
If i <> AcR Then
P2 = M(i, AcC)
    For j = 2 To nCol
            M(i, j) = M(i, j)-P2 * M(AcR, j): 'Debug.
Print i, j, M(i, j), M(i, AcC), M(AcR, j)
    Next j
End If
Next i
'------
ReDim checkC(1 To nCol)
For j = 2 To (nCol-1)
    If M(1, j) > 0 Then
            checkC(j) = M(1, j)
    End If
Next j
    MinC = WorksheetFunction.Max(checkC())
    k = k + 1
Loop
    LinearP = M
Else
    LinearP = "No improvement"
End If
End Function
```

After the first iteration, the output will become Table 5.2. After pivoting on A(1,2), the basic variables are $\{x_1, x_5, x_6\}$.

TABLE 5.2
Table after First Iteration

$-z$	x_1	x_2	x_3	x_4	x_5	x_6	B
1	0	16	12	−20	0	0	−90
0	1	0	0	1	0	0	4
0	0	1	1	−2	1	0	2
0	0	2	1	−2	0	1	8

The procedure is repeated until all coefficients in the first row are non-positive.

TABLE 5.3

Table after the Fourth Iteration: The Procedure Stops since all c Values Are Non-Positive

$-z$	x_1	x_2	x_3	x_4	x_5	x_6	B
1	−4	0	0	0	−8	−4	−154
0	2	0	1	0	2	−1	4
0	0	1	0	0	−1	1	6
0	1	0	0	1	0	0	4

The final output is in Table 5.3. In this table, the three non-basic variables are x_1, x_5 and x_6. Next, the user can work out the optimised solution from the final form. In Table 5.3, the three variables with zero coefficients (x_2, x_3, x_4) are basic variables. The solution corresponds to the combination in which the values of non-basic variables are zero, and the values of basic values are determined by the other rows of the table. The values of basic variables are equal to the corresponding values in column B (since all coefficients of basic variables are 1). For Table 5.3, the B column results in $x_2 = 6$, $x_3 = 4$, and $x_4 = 4$. Since we only need the actual variables in the original problem, the solution is $x_1 = 0$, $x_2 = 6$, $x_3 = 4$. The optimised value of z equals $-B(1) = 154$.

5.3 ADVANCED EXAMPLE: PINCH ANALYSIS OF HEAT EXCHANGER NETWORK

Pinch analysis is a method used in chemical engineering to optimise the design and operation of process systems. It is based on the principle of minimising energy consumption and maximising energy recovery by utilising a heat exchange network. The benefits of pinch analysis include reduced energy consumption, increased energy recovery, improved process efficiency and reduced environmental impact. It is widely used in the chemical process industries, such as petroleum refining, petrochemicals and food processing. Pinch analysis can also be used for the design and optimisation of water networks, where the focus is on minimising water consumption and maximising water reuse within a process. A heat exchanger network connects the hot and cold streams within a processing plant (Serth and Lestina 2014).

EXAMPLE 5.2. DETERMINE A HEAT EXCHANGE NETWORK TO MINIMISE ENERGY CONSUMPTION.

PROBLEM STATEMENT.

Considering a process with two hot streams, two cold streams and two utilities (adapted from example 8.4 (Serth and Lestina 2014). The heat data are tabulated in Table 5.4.

Find the optimal heat exchanger network to minimise the heat requirement.

TABLE 5.4

Heat Data for Heat Exchange Networks

Stream	Type	Ts	Tt ($^{\circ}$C)	Flow	Duty (kW)
H1	Hot	150	60	2.0	180
H2	Hot	90	60	8.0	240
C1	Cool	20	125	2.5	262.5
C2	Cool	20	100	3.0	225
UH1	Hot utility	200	—	—	—
UC1	Cool utility	20	—	—	—

The problem can be converted into a linear optimisation (Pillai and Bandyo-padhyay 2007) with eight flow variables, four equality constraints and two ine-quality constraints. The two utilities, UH and UC, are numbered at heating node 1 and cooling node 1. Other streams are numbered subsequently. Let a_{ij} be the flow from hot stream i to cool stream j. Each non-zero a_{ij} represents a heat exchanger. The required temperatures can be achieved by using the hot utility to heat the two cool streams and the cold utility to cool the two heat streams. This scenario is represented by Figure 5.1(a). In such a setup, the total energy requirement is unnecessarily high. To save energy, the plant operator may want to cross some hot and cool streams. The net energy saving can be achieved by minimising the utility energy.

The objective is to minimise the duty of the cold utility:

$$Z = a_{21} + a_{31} \qquad \text{Eq. 5.6}$$

Since the standard form of linear programming is maximisation, the objective equation is converted to

$$-Z = -a_{21} - a_{31} \qquad \text{Eq. 5.7}$$

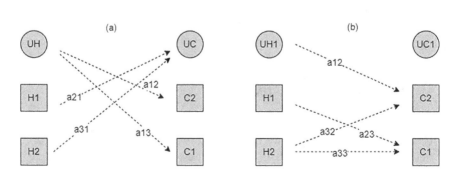

FIGURE 5.1   Heat exchange network. (a) All heating/cooling is done by the two utility streams. (b) Minimisation of heating duty.

Equation forms for meeting energy requirements:

$$a_{12} + a_{22} + a_{32} = 262.5 \qquad \text{Eq. 5.8}$$

$$a_{13} + a_{23} + a_{33} = 225 \qquad \text{Eq. 5.9}$$

$$a_{21} + a_{22} + a_{23} = 180 \qquad \text{Eq. 5.10}$$

$$a_{31} + a_{32} + a_{33} = 240 \qquad \text{Eq. 5.11}$$

Furthermore, one needs to consider the physical constraints. For instance, the problem can include new constraints for temperatures between hot and cool streams. The hot steam H_2 has a very high heat duty, but it cannot heat cool streams to above 90°C (whereas the required temperatures for C1 and C2 are above 90°C). Hence, the heat duty of the exchanger for H2 and C1 is limited to $(90 - 20) \times 2.5$. Similarly, the heat duty between H_2 and C_2 is limited to $(90 - 20) \times 3$. Hence we have two other inequalities:

$$a_{22} < (90-20) \times 2.5 = 140 \qquad \text{Eq. 5.12}$$

$$a_{23} < (90-20) \times 3 = 210 \qquad \text{Eq. 5.13}$$

We have two slack variables, S_{22} and S_{23}, for these equations.

The nice feature of Tables 5.5 is that all coefficients are 1. Yet this isn't easy to work out the basic variables. The problem starts with four basic variables, which will be replaced. However, after the pivoting, only two non-zero (negative) coefficients are in row 1. It is not apparent which variables are non-basic. Hence, the code can be modified to include the entered basic variables, which is AcC. The code inserts three additional rows in the output matrix (Table 5.6).

TABLE 5.5
Heat Data for Heat Exchange Network: Initial Table

Z	a_{12}	a_{13}	a_{21}	a_{22}	a_{23}	a_{31}	a_{32}	a_{33}	S_{12}	S_{22}	B
1				1	1		1	1			420
	1			1			1				262.5
		1			1			1			225
			1	1	1						180
					1	1	1				240
				1					1		175
					1					1	210

TABLE 5.6

Heat Data for Heat Exchange Networks: Final Answer

Z	a_{12}	a_{13}	a_{21}	a_{22}	a_{23}	a_{31}	a_{32}	a_{33}	S_{12}	S_{22}	B
1	0	0	−1	0	0	−1	0	0	0	0	0
0	1	1	−1	0	0	−1	0	−0	0	0	67.5
0	0	1	−1	−1	0	0	0	1	0	0	45
0	−0	−0	1	1	1	0	0	0	−0	−0	180
0	−0	−1	1	1	0	1	1	0	−0	−0	195
0	−0	−0	0	1	0	0	0	0	1	−0	175
0	0	0	−1	−1	0	0	0	0	0	1	30

```
Function LinearP2(A)
Dim checkV(), sols()
ReDim M(1 To nRow + 3, 1 To nCol)
'-----same as function LinearP
'------
M(nRow + 1, 1) = "No. of iter:": M(nRow + 1, 2) = k
ReDim checkV(2 To nCol-1): ReDim sols(2 To nCol-1)
M(nRow + 2, 1) = "Variable": M(nRow + 3, 1) = "Value "
        For i = 2 To nCol-1
            If M(1, i) = 0 Then
                For j = 2 To nRow
                    If M(j, i) = 1 Then
                    checkV(i) = checkV(i) + 1
                    sols(i) = M(j, nCol) 'if
more than 1 positive coef, sols(i) is the last one
                    End If
                Next j
            End If
            If checkV(i) = 1 Then
                M(nRow + 2, i) = "BV": M(nRow +
3, i) = sols(i)
            Else
                M(nRow + 2, i) = "NBV": M(nRow +
3, i) = ""
            End If
        Next i
        LinearP2 = M
Else
            LinearP2 = "No improvement"
End If
End Function
```

TABLE 5.7
Pinch Analysis of Heat Exchange Networks, with Answers

Z	a_{12}	a_{13}	a_{21}	a_{22}	a_{23}	a_{31}	a_{32}	a_{33}	S_{12}	S_{22}	B
1	0	0	−1	0	0	−1	0	0	0	0	0
0	1	1	−1	0	0	−1	0	−0	0	0	67.5
0	0	1	−1	−1	0	0	0	1	0	0	45
0	−0	−0	1	1	1	0	0	0	−0	−0	180
0	−0	−1	1	1	0	1	1	0	−0	−0	195
0	−0	−0	0	1	0	0	0	0	1	−0	175
0	0	0	−1	−1	0	0	0	0	0	1	30
No. of iter:	3	0	0	0	0	0	0	0	0	0	0
Variable	BV	NBV	NBV	NBV	BV	NBV	BV	BV	BV	BV	0
Value	67.5				180		195	45	175	30	0

Code 5.4. To include the variables/answers in the output.

The answers, including the values for basic variables, are given in Table 5.7.

This answer corresponds to the diagram in Figure 5.1(b). From these data, one can calculate other properties of the proposed heat exchange network.

5.4   FURTHER EXAMPLES FOR SELF-PRACTICE

EXAMPLE 5.3.   MAXIMISE A LINEAR FUNCTION WITH A REAL NUMBER.

Using the previous code to solve the following problem:
Maximise the following function:

$$z = 20x_1 + 16x_1 + 12x_3 + 10$$

Subject to:

$$x_1 < 4$$
$$2x_1 + 2x_3 < 10$$
$$2x_1 + 4x_2 + x_3 < 16$$
$$x_3 > -5$$

All variables, except x_3, are non-negative.

Solution

Since x_3 can be both positive and negative, it can be replaced by $x_7 - x_8$ (which are both positive). Thus, a new slack variable is introduced:

$$-x_7 + x_8 + x_9 = 5$$

The previous equation is combined with other equations to form the table.

$-z$	x_1	x_2	x_7	x_8	x_4	x_5	x_6	x_9	B
1	20	16	12	−12					−10
	1	0	0		1	0	0		4
	2		2	−2		1			10
	2	4	1	−1			1		16
			−1	1				1	5

Applying the code will get $x_1 = 4$, $x_2 = 1.75$ and $x_3 = 1$ (from the values of x_7 and x_8).

EXAMPLE 5.3.   MAXIMISE A PROFIT.

A truck is to be used to transport two products to their customers. The volume of the truck is 10,000 L, and the maximum weight that can be carried is 20 t.

- Product A occupies 50 L and weighs 200 kg.
- Product B occupies 80 L and weights 100 kg.
- The profit margins are $5 per unit of Product A and $7 per unit of Product B.

As the truck has limitations on both weight and volume, determine how much of each product to load to maximise the profit.

Solution

Introducing two slack variables, one can generate the following table.

$-z$	A	B	x_1	x_2	B
1	5	7			0
Volume	50	80	1		10000
Weight	200	100	0	1	20000

Applying the previous code, the answers are $B = 90.91$ and $A = 54.55$. Since the numbers of the transported products have to be integers, there will be four combinations around this combination. The best combination is (90, 55).

6 Roots of Systems of Linear Equations

6.1 INTRODUCTION

Systems of linear equations are very common in engineering. For example, one can find them with a series of reactors. However, the most common problems are related to process heat transfer. A system of linear algebraic equations can be solved simultaneously using a direct method such as Gaussian elimination. The Gaussian method consists of two steps: forward elimination and back substitution. Forward elimination reduces all elements in the bottom half to zero (such as in Eq. 6.1). Subsequently, the solution is found from the last variable (x_n to x_1).

$$\begin{bmatrix} a_{11} & a_{12} & a_{13} & \cdots & a_{1n} \\ 0 & a'_{22} & a'_{23} & \cdots & a'_{2n} \\ 0 & 0 & a''_{33} & \cdots & a''_{3n} \\ \vdots & \vdots & \vdots & \cdots & \vdots \\ 0 & 0 & 0 & 0 & a_{nn} \end{bmatrix} \begin{bmatrix} x_1 \\ x_2 \\ x_3 \\ \vdots \\ x_n \end{bmatrix} = \begin{bmatrix} b_1 \\ b'_2 \\ b''_3 \\ \vdots \\ b_n \end{bmatrix}$$

Eq. 6.1

Another direct method is the Gauss–Jordan method. In this method, the final matrix contains nonzero values on the diagonal rather than a triangular matrix. The rows are also normalised by dividing each one by its pivot element. The result is that the identity matrix is formed, where all diagonal coefficients are the value 1. The Gauss–Jordan method requires more work and more operations than Gauss elimination. Gaussian elimination requires order $n^3/3$ multiplications followed by back substitution requiring order n^2 multiplications. Gauss–Jordan elimination requires order $n^3/2$ multiplications (Lindfield and Penny 2019).

The methods can be problematic due to round-off errors and zero on the diagonal. For many problems in process engineering, a successive substitution is more advantageous. In these methods, the initial guesses (a combination of n values of unknowns) are generated and then improved iteratively. These iterative substitutions require an initial guess of the solution. Subsequently, this set of solutions is improved until the errors in all variables are acceptable. The number of iterations of these methods depends on the initial guess and the required error. An "educated" guess can significantly reduce the iterations and multiplications. In addition, the number of iterations can be further reduced by employing a relaxation factor. The VBA codes in this book will only cover the iterative methods.

6.2 SUCCESSIVE SUBSTITUTION METHODS

Two successive substitution methods to improve the solutions are Jacobi and Gauss–Seidel. In both methods, the variables are updated one at a time. In the Jacobi method, all variables in the jth iteration are updated from the values of the $(j-1)^{th}$ iteration. In the Gauss–Seidel, each variable is updated using the latest available values of the other variables. In general, the Gauss–Seidel method provides faster convergence than the Jacobi method.

The substitution formulas for a 4×4 system are presented by four separate equations:

$$x_1^j = \frac{b_1 - a_{12}x_2^{j-1} - a_{13}x_3^{j-1} - a_{14}x_4^{j-1}}{a_{11}}$$

Eq. 6.2a

$$x_2^j = \frac{b_2 - a_{21}x_1^j - a_{23}x_3^{j-1} - a_{24}x_4^{j-1}}{a_{22}}$$

Eq. 6.2b

$$x_3^j = \frac{b_3 - a_{31}x_1^j - a_{32}x_2^j - a_{34}x_4^{j-1}}{a_{33}}$$

Eq. 6.2c

$$x_4^j = \frac{b_4 - a_{41}x_1^j - a_{42}x_2^j - a_{43}x_3^j}{a_{44}}$$

Eq. 6.2.d

Note that the previous equation form is not easy to generalise. The code must contain the four separate equations in each iteration. Furthermore, the form is more complicated for systems with different numbers of equations. Consequently, the formula is re-arranged to apply to larger systems of equations. For instance, the first equation in system 6.2 is re-arranged as:

$$x_1^j = \frac{b_1}{a_{11}} + \frac{a_{11}x_1^{j-1}}{a_{11}} - \frac{a_{11}x_1^{j-1} + a_{12}x_2^{j-1} + a_{13}x_3^{j-1} + a_{14}x_4^{j-1}}{a_{11}}$$

Eq. 6.3

Which can be generalised for all substitutions as:

$$x_i^j = \frac{b_i}{a_{ii}} + \frac{a_{ii}x_i^{j-1}}{a_{ii}} - \frac{\sum a_{ik}x_k^{j-1}}{a_{ii}}$$

Eq. 6.4

While the previous equation looks more complicated than equations in 6.2, the coding is simplified using a second loop (to calculate the last term of the right-hand side). Furthermore, the procedure can be used for any dimension (rather than $n = 4$). The usage of the generalised formula can be seen in the following example codes.

EXAMPLE 6.1. SIMULATING A SERIES OF CONTINUOUS STIRRED-TANK REACTORS.

PROBLEM STATEMENT

The chemical process consists of four continuous stirred-tank reactors (CSTRs) and two recycle streams, as in Figure 6.1. In the reactors, the chemical conversion is given by

$$A \xrightarrow{k} B$$

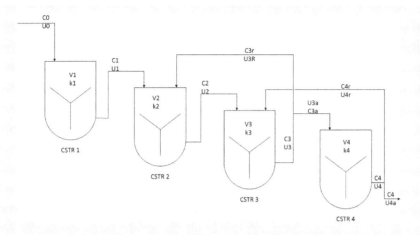

FIGURE 6.1 Series of four CSTRs with two recycle streams.

where k is the rate constant (1/hr), which depends on the operating condition of the reactor.

The volumes and rate constants of the four reactors are tabulated in Table 6.1. The feed conditions are $C_0 = 1$ M, $U_0 = 1000$ L/hr.

Part A. Prepare the system in a matrix form in an Excel spreadsheet.

Part B. Develop code to solve the system using the Gauss–Seidel method.

Part C. Repeat part B using the Jacobi method.

TABLE 6.1
Reactor Size and Conversion Constants

Reactor	1	2	3	4
Vol (L)	1000	1500	200	1000
k (h^{-1})	0.1	0.2	0.4	0.3

Solution

Part A. The problem requires a system of linear equations. At steady state, the mass balances for each reactor are:

$$U_{in} = U_{out} \qquad \text{Eq. 6.5}$$

$$U_{in} C_{in} = U_{out} C_{out} + (kV)C_{out} \qquad \text{Eq. 6.6}$$

where: V is the reactor volume (L)
U_{in}, U_{out} are inflow and outflow, respectively (L/hr)
C_{in}, C_{out} are **A** concentrations in inflow and outflow, respectively (mol/L)
k is the rate constant (1/hr)

TABLE 6.2

Inputs for a System of Linear Equations

	A	B	C	D	E	F
1		Matrix a			$Cguess$	B
2	1100				1	1000
3	1000	−1450	150		1	0
4		1150	−1305.6	75.59866	1	0
5			1075.599	−1375.6	1	0

Consequently, a system of 4×4 equations can be constructed. The elements of the matrices A and B are determined from V, k and corresponding flow rates. In the table, the guess values of all concentrations are 1 M (that is, the value of C_0). The table will be used as input for the VBA functions.

Part B. To solve the previous system, VBA code is developed based on the Gauss–Seidel method. It is noteworthy that the code calls A, B and $Cguess$ as three separate arguments instead of using a single input as a 4×6 table (Table 6.2). The advantage of the arrangement is that the code be used for other dimensions.

Code 6.1. Gauss–Seidel method.

```
Option Base 1
Function GS(a, B, Cguess, tol)
Dim err(10), row, col, n, sum, Cin(10), cnew(10),
MaxErr, k
MaxErr = 1: k = 0: n = a.Rows.Count 'n = number of
rows in A
For row = 1 To n
    Cin(row) = Cguess(row)
Next row
While MaxErr > tol And k < 100
k = k + 1: MaxErr = 0 'increase k after each
iteration
    For row = 1 To n
            sum = 0 'reset sum
            For col = 1 To n
                    sum = sum + a(row, col) * Cin(col) '
            Next col

            cnew(row) = (B(row)-sum + Cin(row) * a(row,
row))/a(row, row) 'calculate Cnew
            err(row) = Abs(cnew(row)-Cin(row)) 'there
will be n errors, the code stops when ALL errors <
tol, or sum of all errors < tol . . .
```

```
                Cin(row) = cnew(row) 'updating values,
cin(1,2,3,4) will be the input of the next iteration
            If err(row) > MaxErr Then
            MaxErr = err(row)
            End If 'calculate MaxErr for each iteration
        Next row
Wend
Cin(n + 1) = k: Cin(n + 2) = MaxErr
GS = Cin
End Function
```

Using the VBA function in Code 6.1, a solution is found after 13 iterations (tol = 10^{-10}).

Part C. For the Jacobi method, the updating loop "Cin(row) = cnew(row)" is separated from the calculating loop. Mathematically, the two methods require the same number of calculations. In this instance, the numbers of iterations are also the same. The function for the Jacobi method is presented in Code 6.2.

Code 6.2. Jacobi method.

```
Function Jacobi(a, B, Cguess, tol)
Dim err(10), row, col, n, sum, Cin(10), cnew(10),
MaxErr, k
MaxErr = 1: k = 0: n = a.Rows.Count
For row = 1 To n
    Cin(row) = Cguess(row)
Next row
While MaxErr > tol And k < 100
k = k + 1: MaxErr = 0 '
    For row = 1 To n
            sum = 0
            For col = 1 To n
                    sum = sum + a(row, col) * Cin(col)
            Next col
                    cnew(row) = (B(row)-sum + Cin(row) *
a(row, row))/a(row, row)
                    err(row) = Abs(cnew(row)-Cin(row))
            If err(row) > MaxErr Then
                    MaxErr = err(row)
            End If
    Next row
    For row = 1 To n
```

```
          Cin(row) = cnew(row) 'updating all values
at the same time
   Next row

Wend
Cin(n + 1) = k: Cin(n + 2) = MaxErr
Jacobi = Cin
End Function
```

Using a *tol* of 10^{-10} and input in Table 6.2, the function in Code 6.1 gives the answers after 25 iterations.

6.3 SUCCESSIVE SUBSTITUTION METHODS WITH A RELAXATION CONSTANT

To enhance the *convergence*, an iterative program can introduce a *relaxation*, in which the value at a particular iteration is made up of a combination of the old value and the newly calculated value:

$$x_i^j = \lambda x_i^j + (1 - \lambda) x_i^{j-1}$$ Eq. 6.7

where λ is a weighting factor assigned a value between 0 and 2.

The relaxation can be classified into three scenarios:

- $0 < \lambda < 1$: under-relaxation
- $\lambda = 1$: no relaxation; the code is the same as the Gauss–Seidel or Jacobi technique
- $1 < \lambda \leq 2$: overrelaxation

EXAMPLE 6.2.   REDUCE THE CALCULATION TIME OF THE GAUSS–SEIDEL METHOD.

PROBLEM STATEMENT

Repeat Part B of Example 6.1 with a relaxation factor.

Solution

The VBA function in Code 6.1 is modified to include a relaxation factor. Using the function in Code 6.3, one can vary λ to see the impact on the number of iterations.

Code 6.3. A Gauss–Seidel method with relaxation.

```
Function GSL(a, b, Cguess, tol, lambda)
Dim err(10), row, col, n, sum, Cin(10), cnew(10),
MaxErr, k
MaxErr = 1: k = 0: n = a.Rows.Count 'n = number of
rows in matrix a
For row = 1 To n
    Cin(row) = Cguess(row)
Next row
While MaxErr > tol And k < 100
k = k + 1: MaxErr = 0 'increase k after each
iteration
    For row = 1 To n
            sum = 0 'reset sum
            For col = 1 To n
                    sum = sum + a(row, col) * Cin(col)
            Next col
            cnew(row) = (b(row)-sum + Cin(row) * a(row,
row))/a(row, row) 'calculate Cnew
            err(row) = Abs(cnew(row)-Cin(row))
            Cin(row) = lambda * cnew(row) + (1-lambda)
* Cin(row)
            If err(row) > MaxErr Then
            MaxErr = err(row)
            End If 'calculate MaxErr for each iteration
    Next row
Wend 'next iteration
Cin(n + 1) = k: Cin(n + 2) = MaxErr
GSL = Cin
End Function
```

The number of iterations can be changed significantly using the VBA function in Code 6.3 with different values of λ. For this particular system, an over-relaxation with $\lambda = 1.1$ can reduce the number of iterations to 11.

6.4   CONCLUDING REMARKS

A system of linear equations with the same number of variables can be solved numerically by many different methods. VBA functions based on substitution methods are simple to develop and can be adapted to many problems. Users can improve the efficiency of the methods. Two key aspects are the formation of coefficients and the selection of the initial guesses. Formation equations can be optimised during the transformation of the mathematical equations from the physical system. The initial guess can also be optimised from the physical characteristics. In some processes, the initial guess for the new operating conditions can be used from previous conditions. These examples will be demonstrated in the next chapter.

6.5 FURTHER EXAMPLES FOR SELF-PRACTICE

EXAMPLE 6.2.   OPTIMISE THE RELAXATION CONSTANT.

Identify the optimal λ between 0 and 3 to minimise the number of iterations of the Jacobi method.

Hints

Re-arrange the function in Code 3.3 so that the output is only the number of iteration $Cn(n + 1)$. Consequently, combine the code with an optimisation method in Chapter 4. The smallest number of iterations should be 8, with λ between 1.0381 and 1.0389. The relations between λ and k can also be plotted in Figure 6.2.

EXAMPLE 6.3.   OPTIMISE THE INITIAL SOLUTION.

In previous solutions, the values for *Cguess* were input by the users. Write the code to generate *Cguess* automatically from *A* and *B*.

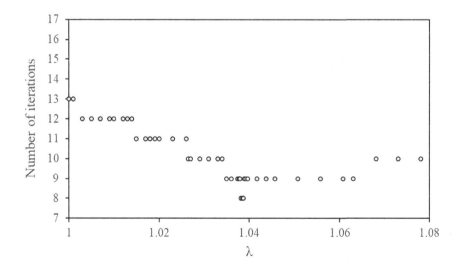

FIGURE 6.2   Relationship between relaxation factor and the number of iterations.

Hints

An obvious solution is $Cg(i) = B(i)/A(i,i)$. If A is a diagonally dominant matrix, these guesses are closer to the final answer than random guesses. The code will need A and B matrices only. Hence, the initial values for the first iteration can be replaced.

```
For row = 1 To n
    Cin(row) = b(row)/a(row, row)
Next row
```

7 Applications of Systems of Linear Equations

7.1 INTRODUCTION

The solution of linear equations can be combined with other methods to solve complex problems. Furthermore, the system of linear equations can be used to impose relations between the various partial derivatives of a multivariable function. Such problems are often found in the process of heat transfer. This chapter will present these problems and solutions.

7.2 INPUT VALIDATION

As can be seen in the previous chapter, the operation will require the dividing of diagonal elements. Hence, it is essential to confirm that the diagonal elements are dominant and non-zero. Consequently, it is important to check the coefficient matrix before proceeding with substitution. To validate if a matrix is a square, one can use the *rows.count* and *column.count* functions. The validation code, such as Code 7.1, can be used in conjunction with other iterations (Code 6.1, 6.2 or 6.3).

Code 7.1. Validating coefficient matrix before applying the Gauss–Seidel algorithm.

```
Function validate(a)
Dim i, n
validate = "Square matrix without zero on diagonal"
n = a.Rows.Count
If n <> a.Columns.Count Then
validate = "Not a square matrix"
Else
For i = 1 To n
If a(i, i) = 0 Then
validate = "Some diagonal elements are zero"
End If
Next i
End If
End Function
```

In practice, these diagonal elements should be dominant, $abs(a(i,i)) > abs(a(i,j))$. Consequently, the sorting procedure may be needed. In the literature, **partial**

 DOI: 10.1201/9781003364511-7

pivoting is used to swap two rows of coefficients when necessary (to place the maximum coefficient of a given column in the proper row). However, as seen in the following examples, some systems already meet the requirement during the formulation process from the physical equations.

7.3 ADVANCED EXAMPLE: A COMBINATION OF BISECTION AND JACOBI METHODS

Once the system of equations has been solved, the code can be combined with other methods to predict other calculations. The following is an example of modelling a series of reactors.

EXAMPLE 7.1. MANIPULATION OF A SERIES OF CSTRS.

PROBLEM STATEMENT

Reconsider the problem in Example 6.1. Adjust recycle stream 2 so that $C_4 = 0.5$ M.

Solution

In this case, a bisection code can be used. All elements of matrices A and B in the previous example have to be calculated within the VBA code. Please note that only two elements in matrix A depend on U4R. Subsequently, the bisection method can be applied. First, the data are entered into the specified cells of an Excel spreadsheet (Table 7.1)

The first section of the VBA code is used to generate the matrices A, B and $Cguess$ before running the Gauss–Seidel or Jacobi method. The elements of A and B are calculated according to the mass balances of the reactors. The output of the VBA code

TABLE 7.1
Reactor Sizes, Constants, and Flow Conditions

	A	B	C
1	Reactor volume and kinetic constant		
2	Reactor	Vol (L)	k (h^{-1})
3	1	1000	0.1
4	2	1500	0.2
5	3	200	0.4
6	4	1000	0.3
7	Flow conditions		
8	C_{A0} (mol/L)	1	
9	U_0 (L/hr)	1000	
10	U_{3R} (L/hr)	150	

is the concentration of stream 4 (*C4*). Consequently, the bisection code is used to determine U4R so that *C4* = 0.5 M. This function is shown in Code 7.2.

Code 7.2. Modifying Gauss–Seidel code for optimisation.

```
Option Base 1
Function GSv(U4r, tol)
Dim err(100), row, col, n, sum, Cin(100), cnew(100), MaxErr, k
Dim a(4, 4), Cguess(4), B(4), V(4), kin(4), Ca0, U0,
U3rV(1) = Range("B3").Value: kin(1) = Range("C3").Value
V(2) = Range("B4").Value: kin(2) = Range("C4").Value
V(3) = Range("B5").Value: kin(3) = Range("C5").Value
V(4) = Range("B6").Value: kin(4) = Range("C6").Value

Ca0 = Range("B8").Value
U0 = Range("B9").Value
U3r = Range("B10").Value

a(1, 1) = U0 + V(1) * kin(1)
a(2, 1) = U0: a(2, 2) = -(U0 + U3r)-V(2) * kin(2): a(2,
3) = U3r
a(3, 2) = U0 + U3r: a(3, 3) = -(U0 + U3r + U4r)-V(3) *
kin(3): a(3, 4) = U4r
a(4, 3) = U0 + U4r: a(4, 4) = -(U0 + U4r)-V(4) * kin(4)
B(1) = U0 * Ca0
For k = 1 To 4
Cguess(k) = Ca0
Next k
n = 4
'****** repeat Jacobi or Gauss-Seidel code (Code 6.1.)
'******
GSv = Cin(4)
End Function

Function BisectionX(Lo, Up, tol)
Dim Mi, c1, c3, i, n
    n = Abs(Up-Lo) / 0.01
'*****************************************
c1 = GSv(Lo, tol)-0.5
For i = 1 To n
    Mi = (Lo + Up) / 2
    c3 = GSv(Mi, tol)-0.5
If c1 * c3 < 0 Then
    Up = Mi
Else
```

```
      Lo = Mi: c1 = c3
 End If
 Next i
 '*****************************************
 BisectionX = Mi
 End Function
```

Using the previous code with a tolerance of 10^{-4}, the required U4R is determined at 75.2794 L/hr and $C_4 = 0.50000$ M. With a tolerance of 10^{-6}, the optimised U4R is determined at 75.5018 L/hr and $C_4 = 0.50000$ M. The sensitivity of the correlation between C_4 and U4R is a key advantage of employing recycle streams in reaction engineering.

7.4 ADVANCED EXAMPLE: 2-DIMENSIONAL STEADY-STATE HEAT TRANSFER

In 2-dimensional steady-state heat transfer, the temperature distribution in a solid/fluid object is calculated when the heat transfer rate is constant over time. This type of heat transfer is important in various engineering applications, such as designing heating and cooling systems for buildings and calculating the temperature distribution in electronic devices.

The heat transfer equation for 2-dimensional steady-state heat transfer is given by:

$$\nabla^2 T = -Q/k \qquad \text{Eq. 7.1}$$

where T is the temperature at a point in the body, Q is the heat transfer rate per unit volume and k is the material's thermal conductivity.

In Eq. 7.1, $\nabla^2 T = -Q/k$ represents the Laplacian of T, which is the sum of the second partial derivatives of T with respect to the x and y coordinates. The equation can be solved using numerical methods, such as finite difference, finite element or boundary element methods. These methods discretise the object or fluid into a grid of points, and the temperature at each point is calculated based on the temperature and heat flux at neighbouring points. Since the temperatures of all points are interlinked, a system of equations must be solved simultaneously.

EXAMPLE 7.2. STEADY-STATE TEMPERATURE PROFILE (WHITE 2002).

PROBLEM STATEMENT

A metal plate is exposed to the surrounding environment, as shown in Figure 7.1.

Calculate the temperature of the internal nodes.

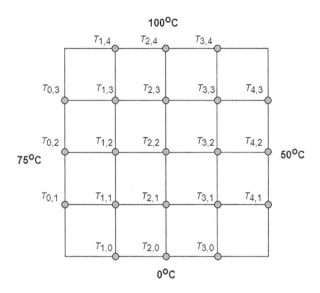

FIGURE 7.1   The discretisation of the 2-dimensional heating plate.

Solution

For 2-dimensional steady-state heat problems, there is no heat accumulation at any point. The temperature of the boundary nodes is the same as the surrounding environment. Therefore, the values of temperature at the four corners are ignored. For any interior nodes, the temperature is given the balance of the four surrounding nodes.

For instance, the temperature of node (1,1) is given by:

$$4T_{1,1} = T_{0,1} + T_{1,0} + T_{2,1} + T_{1,2} \qquad \text{Eq. 7.2}$$

With boundary nodes having constant temperatures, a system of nine equations is established:

$$
\begin{bmatrix}
-4 & 1 & 0 & 1 & 0 & 0 & 0 & 0 & 0 \\
1 & -4 & 1 & 0 & 1 & 0 & 0 & 0 & 0 \\
0 & 1 & -4 & 0 & 0 & 1 & 0 & 0 & 0 \\
1 & 0 & 0 & -4 & 1 & 0 & 1 & 0 & 0 \\
0 & 1 & 0 & 1 & -4 & 1 & 0 & 1 & 0 \\
0 & 0 & 1 & 0 & 1 & -4 & 0 & 0 & 1 \\
0 & 0 & 0 & 1 & 0 & 0 & -4 & 1 & 0 \\
0 & 0 & 0 & 0 & 1 & 0 & 1 & -4 & 1 \\
0 & 0 & 0 & 0 & 0 & 1 & 0 & 1 & -4
\end{bmatrix}
\begin{bmatrix}
T_{1,1} \\ T_{2,1} \\ T_{3,1} \\ T_{1,2} \\ T_{2,2} \\ T_{3,2} \\ T_{1,3} \\ T_{2,3} \\ T_{3,3}
\end{bmatrix}
=
\begin{bmatrix}
-75 \\ 0 \\ -50 \\ -75 \\ 0 \\ -50 \\ -175 \\ -100 \\ -150
\end{bmatrix}
\qquad \text{Eq. 7.3}
$$

One can use the previous VBA functions, such as Code 6.1 (Gauss–Seidel method), to solve the previous system. A small modification can be done, as shown in Code

7.3, to have the number of outputs exactly the same as the number of equations and displaced in the vertical direction.

Code 7.3. Modifying Gauss–Seidel code for 2-dimensional steady-state heat transfer.

```
Function GSv(a,b,Cguess,tol)
Dim output()
'-----same as Code 6-1
'------
ReDim output(n, 1)
For row = 1 To n
output(row, 1) = Cin(row)
Next row
GSv = output
```

Using the VBA function in Code 7.3, the answer will be given in Table 7.2. The results are also presented graphically in Figure 7.2.

TABLE 7.2

The Temperature of Internal Nodes

T11	42.76562
T21	33.16739
T31	33.8828
T12	63.07809
T22	56.15845
T32	52.40958
T13	78.52565
T23	76.07029
T33	69.61997

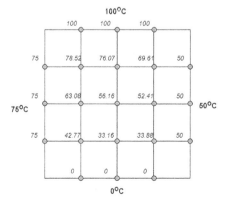

FIGURE 7.2 Steady-state temperature distribution.

7.5   ADVANCED EXAMPLE: UNSTEADY-STATE HEAT TRANSFER AFTER MICROWAVE HEATING

The unsteady-state heat conduction is a partial differential equation. The equation is also applied to molecule diffusion. The numerical methods can be classified as implicit, explicit and Crank–Nicholson methods (Chapra and Canale 2010). In short, the implicit method predicts the new values in the time domain from the previous values. The Crank–Nicholson method can be considered an average of the implicit and explicit methods. The Crank–Nicholson method is unconditionally stable. In terms of numerical calculation, both explicit and Crank–Nicholson methods require solving a system of linear equations.

EXAMPLE 7.3.   COOLING OF AN AQUEOUS SOLUTION AFTER MICROWAVE HEATING.

PROBLEM STATEMENT

Consider the layer of water of 5 cm, which was microwaved to 80°C. The layer was cooled by exposure to an oil layer at a constant temperature of 23°C. Assume that heat loss to the container wall is negligible. Water thermal diffusivity is given (10^{-6} m²/s):

$$k = 0.1319 + 6.1231 \times 10^{-4} \times T - 2.5966 \times 10^{-6} \times T^2$$

Part A. Calculate the temperature profile in the vertical direction, from the oil surface to 1.2 cm depth, after 1, 2, 5 and 10 mins.
Part B. Calculate the temperature at the points of 1.2 mm from the surface and compare it with experimental data (Shibata et al. 2018).

Solution

The heat diffusion equation is given by:

$$\frac{\partial T}{\partial t} = k \frac{\partial^2 T}{\partial x^2}$$

Eq. 7.3

where k is the thermal diffusivity of water, which varies with temperature.

The vertical domain of the water body is divided into n nodes. Node 1 is at the 1.2 cm depth, and node n is the water surface. At $t = 0$, all temperatures, except node n, equal 80°C.

The dimensionless parameter, λ, is given:

$$\lambda = k \frac{\Delta t}{\Delta x^2}$$

Eq. 7.4

For internal nodes (from 2 to $n-1$), the heat diffusion is discretised as:

$$-\lambda T_{i-1}^{l+1} + 2(1+\lambda)T_i^{l+1} - \lambda T_{i+1}^{l+1} = \lambda T_{i-1}^l + 2(1+\lambda)T_i^l + \lambda T_{i+1}^l \qquad \text{Eq. 7.5}$$

The bottom node is modelled as an insulated boundary condition, $dT/dx(x = xb) = 0$. Imagine there is a frictional node outside the domain $T(t,-1)$; the insulated condition indicates that $T_{-1}^{l+1} = T_1^{l+1}$. Consequently, the equation for the first node is:

$$-\lambda T_1^{l+1} + 2(1+\lambda)T_0^{l+1} - \lambda T_1^{l+1} = \lambda T_1^l + 2(1+\lambda)T_0^l + \lambda T_1^l \qquad \text{Eq. 7.6}$$

Finally, the equation for node n:

$$T_n^{l+1} = 23 \qquad \text{Eq. 7.7}$$

The previous system is arranged into a system of n equations. The $n \times n$ matrix has non-zero values in the next diagonal and is often known as the tridiagonal matrix. The procedure is repeated m times to get the answers in m seconds (Code 7.2). This code consists of a separate function, SS(T, delx, delt), to calculate the value of λ at each node temperature.

Code 7.4. One-dimensional heat transfer after microwave heating.

```
Option Base 1
Function GS _ 3t(n, m, Tpre, T0, delt, delx, tol) 'C-N,
n is number of nodes
Dim a(), B(), err(), row, col, sum, T(), Tin(),
Tnew(),3t _ t(n, m, Tpre, T0, delt, delx, tol) 'C-N, n
is number of nodes MaxErr, k, i, s
ReDim a(1 To n, 1 To n), B(1 To n), err(1 To n), Tin(1
To n + 2), Tnew(1 To n), T(1 To n + 2, 1 To m)

'--- To is the surface temperature- constant value
For row = 1 To n-1
T(row, 1) = Tpre: Tin(row) = Tpre
Next row
Tin(n) = T0: T(n, 1) = T0

For i = 2 To m
MaxErr = 1: k = 1
'------------
While MaxErr > tol And k < 1000
k = k + 1: MaxErr = 0
'--------------
a(1, 1) = 2 + 2 * SS(Tin(1), delx, delt): a(1, 2) = -2
* SS(Tin(1), delx, delt)
```

```
B(1) = 2 * (1-SS(T(1, i-1), delx, delt)) * T(1, i-1) + 2
* SS(T(1, i-1), delx, delt) * T(2, i-1)
For row = 2 To n-1
    s = SS(Tin(row), delx, delt)
    a(row, row) = 2 + 2 * s: a(row, row-1) = -s:
a(row, row + 1) = -s
    B(row) = 2 * (1-s) * T(row, i-1) + s * T(row-1,
i-1) + s * T(row + 1, i-1)
Next row
a(n, n) = 1: B(n) = T0
'------- . . . . . Gauss-Seidel (code 6.1) or Jacobi
(code 6.2)
    For row = 1 To n
            sum = 0 'reset sum
            For col = 1 To n
                    sum = sum + a(row, col) * Tin(col)
            Next col
                    Tnew(row) = (B(row)-sum + Tin(row) *
a(row, row))/a(row, row)
                    err(row) = Abs(Tnew(row)-Tin(row))
                    Tin(row) = Tnew(row)
            If err(row) > MaxErr Then
                    MaxErr = err(row)
            End If
    Next row
Wend
Tin(n + 1) = k: Tin(n + 2) = MaxErr
For row = 1 To n + 2
    T(row, i) = Tin(row)
Next row
Next i
For i = 1 To n '4 profiles at 60, 120, 300 and 600
seconds
    Output(i, 1) = T(i, 60): Output(i, 2) = T(i, 120)
    Output(i, 3) = T(i, 300): Output(i, 4) = T(i, 600)
Next i
GS _ 3t = Output
End Function
Function SS(T, delx, delt)
Dim k
k = 0.1319 + 6.1231 * 10 ^ (-4) * T-2.5966 * 10 ^ (-6)
* T ^ 2
SS = k * delt/delx ^ 2
End Function
```

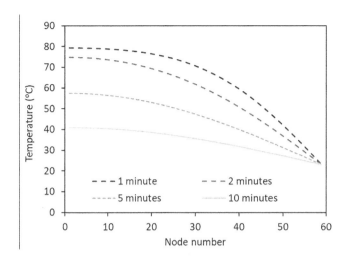

FIGURE 7.3 Temperature profiles near water surface after microwave.

It should be noted the boundary condition is mathematically the same as an insulated boundary condition. The value of λ is calculated based on the temperature of T_{l+1}. Alternatively, λ can be calculated from the average of T_l and T_{l+1}. However, such an improvement makes the code more complicated without significantly improving accuracy. The code will give temperature values in an Excel spreadsheet and the plot in Figure 7.3.

For Part B, the output is modified to give the temperate profile of node 1 (Code 7.5). The results are plotted in Figure 7.4.

Code 7.5. One-dimensional heat transfer after microwave heating, producing a temperature at 12 mm depth (the main part of the code is the same as in Code 7.4).

```
Function GS _
Dim a(), B(), err(), row, col, sum, T(), Tin(), Tnew(),
Output(), MaxErr, k, i, s
'Dim Cout(1, 1000)
ReDim a(1 To n, 1 To n): ReDim Output(1 To m, 1 To 1)
'------

'-------
For i = 1 To m
Output(i, 1) = T(1, i)
Next i

GS _ 3t _ t = Output 'ouput as array (cells are in hor-
izontal . . .)
End Function
```

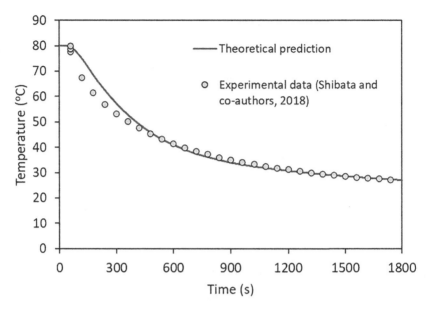

FIGURE 7.4 The temperature at 12 mm from the surface; experimental data were obtained from Shibata and coauthors (2018).

8 Solving Systems of Non-Linear Equations

8.1 INTRODUCTION

A system of non-linear equations is a set of equations where at least one is non-linear, that is, contains a non-linear function such as a polynomial, trigonometric or exponential function. Unlike linear equations, these equations cannot be solved using simple algebraic techniques, and numerical methods must be used to obtain solutions. The chapter presents several methods to solve systems of non-linear equations. While some of the methods can be generalised, the success of a particular method will depend on the nature of the equations and initial guesses. Furthermore, it should be noted that many systems cannot be solved. Applying a numerical code to a general system could lead to unfruitful outcomes.

8.2 NEWTON'S AND SECANT METHODS

While the number of equations and unknowns don't have to be the same, mathematical systems require zero degrees of freedom to have a unique solution. The necessary condition of the unique solution is the number of equations being equal to the number of unknowns. Newton's method in Chapter 2 can be extended for such a system by employing an $m \times m$ Jacobian matrix, where m is the number of equations/unknowns. The Jacobian matrix is the matrix in which all elements are its first-order partial derivatives. The iterations will calculate the next set of variables from the Jacobian of the previous set. The formula is detailed in the following for a 2×2 system.

For example, consider a system of two equations and two unknowns:

$$u(x, y) = 0 \qquad \text{Eq. 8.1a}$$

$$v(x, y) = 0 \qquad \text{Eq. 8.1b}$$

where x and y are independent variables.

The iteration for this 2×2 system is given by:

$$x_{i+1} = x_i - \frac{u_i \dfrac{\partial v_i}{\partial y} - v_i \dfrac{\partial u_i}{\partial y}}{\dfrac{\partial u_i}{\partial x} \dfrac{\partial v_i}{\partial y} - \dfrac{\partial u_i}{\partial y} \dfrac{\partial v_i}{\partial x}} \qquad \text{Eq. 8.2a}$$

DOI: 10.1201/9781003364511-8

$$y_{i+1} = y_i - \frac{v_i \dfrac{\partial u_i}{\partial x} - u_i \dfrac{\partial v_i}{\partial x}}{\dfrac{\partial u_i}{\partial x}\dfrac{\partial v_i}{\partial y} - \dfrac{\partial u_i}{\partial y}\dfrac{\partial v_i}{\partial x}} \qquad \text{Eq. 8.2b}$$

where the partial derivatives are evaluated at the values of x_i and y_i.

An analog to the secant method in Chapter 2 can be used to avoid differential equations. In this case, two different step sizes, *delx* and *dely*, can be applied. VBA code was developed in the following example to calculate the binary adsorption at the air/water surface (Le et al. 2012). It should be noted that the step sizes (*del1* and *del2*) and initial values need to be selected appropriately. If these steps are too large, the solution can be unstable.

EXAMPLE 8.1.   SOLVING TWO NON-LINEAR EQUATIONS BY THE SECANT METHOD.

PROBLEM STATEMENT

The co-adsorption of two surfactants can be modelled by two equations:

$$K_1 C_1 = \frac{\theta_1}{(1-\theta_1-\theta_2)^{r_1}} exp\left\{-2A_1\theta_1 - 2A_{12}\theta_2 + (1-r_1)\left(A_1\theta_1^2 + A_2\theta_2^2 + 2A_{12}\theta_1\theta_2\right)\right\} \quad \text{Eq. 8.3}$$

$$K_2 C_2 = \frac{\theta_2}{(1-\theta_1-\theta_2)^{r_2}} exp\left\{-2A_2\theta_2 - 2A_{12}\theta_1 + (1-r_2)\left(A_1\theta_1^2 + A_2\theta_2^2 + 2A_{12}\theta_1\theta_2\right)\right\} \quad \text{Eq. 8.4}$$

where C_i and Γ_i are the bulk and surface concentrations of surfactants ($i = 1$, 2), respectively, $\theta_i = \Gamma_i / \Gamma_{m,i}$ is the monolayer coverage. $\Gamma_{m,i}$, K_i and A_i are the adsorption constants of the surfactants, which are predetermined individually. A_{12} is a parameter accounting for the cross-interaction between adsorbed molecules of the two surfactants.

The two surface area fractions are defined as: $r_i = 1/\omega\Gamma_{m,i}$

where the mean molar area, ω, is defined as $\omega = \dfrac{\dfrac{\Gamma_i}{\Gamma_{m,i}} + \dfrac{\Gamma_2}{\Gamma_{m,2}}}{\Gamma_1 + \Gamma_2}$

From the binary adsorption, the surface tension (γ) is given as:

$$\gamma = 72 - \frac{RT}{\omega}\ln\left(1-\theta_1-\theta_2\right)A_1\theta_1^2 + A_2\theta_2^2 + 2A_{12}\theta_1\theta_2 \qquad \text{Eq. 8.5}$$

where R and T are the gas constant and temperature, respectively.

For a particular system of surfactant mixture, the parameters are given in Table 8.1. Calculate the surface tension of this mixture at $25°C$ when $c_1 = 9.6\times10^{-3}$ M and $c_1 = 1.7\times10^{-4}$ M.

TABLE 8.1

Parameters for Binary Adsorption of Two Surfactants

	A	B	C	D	E	F	G
1	$\Gamma_{m,1}$ (mol/m^2)	$\Gamma_{m,2}$ (mol/m^2)	K_1 (M^{-1})	K_1 (M^{-1})	A1	A2	A12
2	1.5×10^{-5}	1.1×10^{-5}	40.1	2596	0	0.2	−0.963

Solution

The problem requires solving two equations simultaneously. These two equations are not in an explicit form. However, the two functions are the same, with interchanging roles between **c1** and **c2**. Hence, a common VBA function, f457, can be used.

Code 8.1. Secant method for a system of two non-linear equations.

```
Function jacobian457(c1, c2, k1, k2, w1, w2, a1, a2,
a12, theta1, theta2)
Dim i, Err, del1, del2, u, v, dudx, dudy, dvdx, dvdy, a(0, 3)
Err = 1: del1 = theta1 / 200: del2 = theta2 /200: i = 0

While Err > (0.0000000001) and i < 1000
u = f457(theta1, theta2, w1, w2, c1, b1, a1, a2, a12)
v = f457(theta2, theta1, w2, w1, c2, b2, a2, a1, a12)

dudx = (u-f457(theta1-del1, theta2, w1, w2, c1, k1, a1,
a2, a12))/del1 'd(f1)/d(theta1)
dudy = (u-f457(theta1, theta2-del2, w1, w2, c1, k1,
a1, a2, a12))/del2
dvdx = (v-f457(theta2, theta1-del1, w2, w1, c2, k2,
a2, a1, a12))/del1
dvdy = (v-f457(theta2-del2, theta1, w2, w1, c2, k2,
a2, a1, a12))/del2

theta1 = theta1-(u * dvdy-v * dudy)/(dudx * dvdy-
dudy * dvdx)
theta2 = theta2-(v * dudx-u * dvdx)/(dudx * dvdy-
dudy * dvdx)

Err = u ^ 2 + v ^ 2: i = i + 1
Wend
a(0, 0) = theta1: a(0, 1) = theta2: a(0, 2) = i:
a(0, 3) = (theta1 + theta2)/(theta1 / w1 + theta2 / w2)
jacobian457 = a
End Function
```

```
Function f457(thetai, thetaj, wi, wj, ci, ki, ai, aj,
aij)
Dim w, ni, ff
w = (thetai + thetaj)/(thetai/wi + thetaj/wj)
ni = wi/w
ff = (thetai/(1-thetai-thetaj) ^ ni) * Exp(-2 * ai *
thetai-2 * aij * thetaj)
ff = ff * Exp((1-ni) * (ai * thetai ^ 2 + aj * thetaj
^ 2 + 2 * aij * thetai * thetaj))
f457 = ff-ci * ki
End Function
```

Using theta1 = 0.282 and theta2 = 0.321, the solution is obtained after six iterations: theta1 = 0.183, theta2 = 0.192 and ω = 77271 (m²/mol). From Eq. 8.5, the surface tension is calculated as 55.0 mN/m.

8.3　ADVANCED EXAMPLE: TEAR ANALYSIS WITH TWO INDEPENDENT UNKNOWNS

In the previous problem, the two equations are symmetric; thus, the same VBA function can be used for evaluation. In the following examples, the two equations are different. A similar approach is applicable.

EXAMPLE 8.2.　TEAR ANALYSIS WITH TWO INDEPENDENT VARIABLES.

PROBLEM STATEMENT

Ethyl chloride (CH_3CH_2Cl) can be manufactured by an exothermic reaction of hydrogen chloride (HCl) with ethylene ($CH_2=CH_2$):

$$C_2H_4 + HCl \rightarrow C_2H_5Cl$$

The process can be simplified in the following diagram, adapted from example 10.2–3 (Felder and Rousseau 2005). The fresh feed contains 100 kgmol/h of HCl and C_2H_4 mixture. The fresh feed is mixed with the recycle stream before entering the reactor. The single-pass ethylene conversion is 90%. After the reactor, the stream is separated into Product (containing C_2H_5Cl only) and Stream S-4 (containing C_2H_4 and HCl). Stream S4 is split into Waste and Recycle streams. Streams S4, Waste and Recycle have the same composition. The ethylene molar fraction in the feed is 52%, and the waste stream has a flow rate of 8 kgmol/h.

Find the molar flows for two species in the recycle stream.

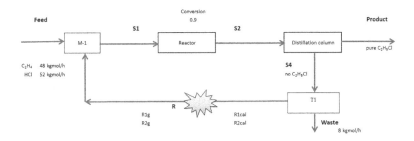

FIGURE 8.1 Acetylene production with a Recycle stream and two unknowns.

Solution

The flowsheet is the same as in Example 3.2. For example, if the molar fraction is 50%, the solution is the same as in Example 3.2 and can be solved by a bisection code. In this case, however, since the molar fractions of the feed are not equal, the solution will have unequal flows in the Recycle stream (Figure 8.1).

For any combination of the R1 and R2, the material balances allow the calculation of the tear stream (R1cal and R2cal). Consequently, the program needs to solve the following equations simultaneously:

$$dif1(R1g, R2g) = R1g - R1cal \qquad \text{Eq. 8.6}$$

$$dif2(R1g, R2g) = R2g - R2cal \qquad \text{Eq. 8.7}$$

Hence, two VBA functions are developed correspondingly.

Code 8.2. Two primary functions for tear analysis.

```
Function difR1(Rg1, Rg2, feed, fraction, Conv, W)
Dim S1(4), S2(4), S4(4), Pro(4), Waste(4), R(4), Rt(4)
Dim ratio

R(1) = Rg1: R(2) = Rg2 'tear stream
S1(1) - feed * fraction ı R(1): S1(2) = feed * (1-frac-
tion) + R(2)
S2(1) = S1(1)-S1(1) * Conv: S2(2) = S1(2)-S1(1) * Conv:
S2(3) = S1(2) * Conv
S4(1) = S2(1): S4(2) = S2(2)
ratio = S4(1)/(S4(1) + S4(2))
Rt(1) = S4(1)-W * ratio: Rt(2) = S4(2)-W * (1-ratio):

difR1 = Rt(1)-R(1)
End Function


Function difR2(Rg1, Rg2, feed, fraction, Conv, W)
Dim S1(4), S2(4), S4(4), Pro(4), Waste(4), R(4), Rt(4)
```

```
Dim ratio
R(1) = Rg1: R(2) = Rg2 'tear stream
S1(1) = feed * fraction + R(1): S1(2) = feed * (1-frac-
tion) + R(2)
S2(1) = S1(1)-S1(1) * Conv: S2(2) = S1(2)-S1(1) * Conv:
S2(3) = S1(2) * Conv
S4(1) = S2(1): S4(2) = S2(2)
ratio = S4(1)/(S4(1) + S4(2))
Rt(1) = S4(1)-W * ratio: Rt(2) = S4(2)-W * (1-ratio):
difR2 = Rt(2)-R(2)
End Function
```

It should be noted that these functions include the same calculation until the last line. With these two functions, the four elements of a Jacobian matrix can be obtained and used to calculate the next set of R1 and R2. The iteration is repeated until the total error is sufficiently small.

Code 8.3. System of two non-linear equations (in combination with tear analysis).

```
Function jacobian(R1i, R2i, feed, fraction, Conv, W)
Dim i, Err, del1, del2, u, v, dudx, dudy, dvdx, dvdy, a(0, 2)

Err = 1: del1 = 0.001: del2 = 0.001: i = 0

While Err > (0.000000001) And i < 100
u = difR1(R1i, R2i, feed, fraction, Conv, W)
v = difR2(R1i, R2i, feed, fraction, Conv, W)

dudx = (u-difR1(R1i-del1, R2i, feed, fraction, Conv,
W))/del1
dudy = (u-difR1(R1i, R2i-del2, feed, fraction, Conv,
W))/del2
dvdx = (v-difR2(R1i-del1, R2i, feed, fraction, Conv,
W))/del1
dvdy = (v-difR2(R1i, R2i-del2, feed, fraction, Conv,
W))/del2

R1i = R1i-(u * dvdy-v * dudy)/(dudx * dvdy-dudy * dvdx)
R2i = R2i-(v * dudx-u * dvdx)/(dudx * dvdy-dudy * dvdx)
Err = u ^ 2 + v ^ 2: i = i + 1
Wend
a(0, 0) = R1i: a(0, 1) = R2i: a(0, 2) = i
jacobian = a
End Function
```

With initial guesses of (5,5), the code gives R1=3.111 kgmol/h and R2=9.333 kgmol/hr after five iterations. If the fraction is changed to 0.5, the code should give the same answer as in Example 3.2. While the method gives a quick solution for this case, extending it to more variables (for example, three or four species in the tear stream) will be difficult. Therefore, in the next section, an empirical search method is introduced.

8.4 ADVANCED EXAMPLE: TEAR ANALYSIS WITH STOCHASTIC OPTIMISATION

Stochastic optimisation is an algorithm that uses random samples or simulations to optimise an objective function. In contrast to deterministic optimisation algorithms, which use exact formulas and methods to optimise an objective function, stochastic optimisation algorithms allow randomness in the prediction. Stochastic optimisation algorithms typically use an iterative procedure to find the optimal solution. At each iteration, the algorithm samples a subset of the data or a random point in the search space and uses it to update the current estimate of the optimal solution. The algorithm then repeats the process until a convergence criterion is met. Stochastic optimisation methods rely on random variables. VBA can utilise the random function (RAND) for such a purpose.

Harmony search is a stochastic optimisation (Ingram and Zhang 2010). It is inspired by the process of musicians improvising music together to produce a harmonious melody. In harmony search, a potential solution to the optimisation problem is represented as "harmony," which consists of a set of values for the decision variables. The algorithm starts with an initial population of randomly generated harmonies. Then each iteration creates a new harmony by blending existing harmonies in the population. The blending is done randomly, aiming to produce a new harmony with a better objective function value than the existing ones. The blending process in harmony search involves three options: memory consideration, pitch adjustment and randomisation. In each iteration, one of these three options is selected randomly. The selection and calculation of these options are explained in the following example.

EXAMPLE 8.3. HARMONY SEARCH WITH TWO INDEPENDENT VARIABLES.

Problem statement

Repeat Example 8.2 using the harmony search method.

Solution

In this case, the previous example is used to demonstrate the method. Instead of solving two equations simultaneously, the harmony search aims to minimise the sum of squares of difference:

$$dif f2 = \left(R2g - R2cal\right)^2 + \left(R1g - R1cal\right)^2 \qquad \text{Eq. 8.8}$$

The function in the previous equation is a total difference between guessed and calculated values of all unknowns in the tear stream. The function is evaluated in Code 8.4, using the same calculation steps in Code 8.2.

Code 8.4. Function to calculate the sum of squares of differences (with tear analysis)

```
Function diffR2(Rg1, Rg2, feed, fraction, Conv, W)
Dim S1(4), S2(4), S4(4), Pro(4), Waste(4), R(4), Rt(4)
Dim ratio
R(1) = Rg1: R(2) = Rg2 'tear stream
S1(1) = feed * fraction + R(1): S1(2) = feed * (1-fraction)
+ R(2)
S2(1) = S1(1)-S1(1) * Conv: S2(2) = S1(2)-S1(1) * Conv:
S2(3) = S1(2) * Conv 'we don't need S(3) for calc
S4(1) = S2(1): S4(2) = S2(2): ratio = S4(1)/(S4(1) +
S4(2))
Rt(1) = S4(1)-W * ratio: Rt(2) = S4(2)-W * (1-ratio):
diffR2 = (Rt(1)-R(1)) ^ 2 + (Rt(2)-R(2)) ^ 2
End Function
```

Consequently, the harmony search can be applied (Codes 8.5 to 8.6). The step-by-step is explained in the following. First, a set of 100 (HMS) combinations of $x1$ and $x2$ are generated randomly. It should be noted that the VBA function RND has a small range of randomness. Therefore, one can improve the process by adding the function RANDOMIZE before using RND. Furthermore, the code can indirectly use the Excel spreadsheet function by typing RAND() into a cell and then EVALUATE that cell. Next, the objective functions of these combinations are evaluated and stored in the HM array (Code 8.4). The code has three key parameters: HMS (Harmony Memory Size), HMS $\geq$ 1; HMCR (Harmony Memory Considering Rate), $0 \leq$ HMCR $\leq$ 1; and PAR (Pitch Adjusting Rate), $0 \leq$ PAR $\leq$ 1 (Ingram and Zhang 2010).

Code 8.5. Harmony search with two variables (the function diffR2 is defined in Code 8.4). Section 1: generate the initial sets of solutions.

```
Function harmony2(LowerR, UpperR, feed, fraction,
Conv, W)
Dim limit(5, 3), x(5), ND
Dim HM(500, 3), HMS, HMCR, PAR, MaxIte
Dim d1, sol, i, j, iter, rand
Dim hmax, hmax _ n, hmin, hmin _ n, out(4, 4)

ND = 2 'number of decision variables

'lower limit, upper limit & # of initial segments
```

```
limit(1, 1) = LowerR: limit(1, 2) = UpperR: limit(1,
3) = 1000
limit(2, 1) = LowerR: limit(2, 2) = UpperR: limit(2,
3) = 1000

HMS = 100: HMCR = 0.9: PAR = 0.6: MaxIte = 100000
'--------generate the 1st sets of 100 combinations
For i = 1 To HMS
    For j = 1 To ND
          x(j) = limit(j, 1) + (limit(j, 2)-limit(j, 1))
* Rnd 'generate random values of x1 and x2 between
lower and upper limits
          HM(i, j) = x(j)
    Next j
    HM(i, 3) = diffR2(x(1), x(2), feed, fraction, Conv, W)
'evaluate the function
Next i
'-------------------
```

Subsequently, the minimum and maximum values within HM are obtained in Code 8.6.

Code 8.6. Harmony search with two variables. Section 2: find the maximum and minimum values from the initial sets.

```
hmax _ n = 1: hmax = HM(1, 3)
For i = 2 To HMS
    If HM(i, 3) > hmax Then
          hmax _ n = i
          hmax = HM(i, 3)
    End If
Next i

hmin _ n = 1: hmin = HM(1, 3)
For i = 2 To HMS
    If HM(i, 3) < hmin Then
          hmin _ n = i: hmin = HM(i, 3)
    End If
Next i
out(1, 1) = "Min from the initial sets": out(1,
2) = "Max from the initial sets"
out(2, 1) = hmin: out(2, 2) = hmax
'---------------
```

In the next step, a new combination is obtained randomly. A random value is generated and compared with the two limits, HMCR and PAR. Depending on the value of RND, either one of three operations will be carried out:

- If RND > HMCR: a new set will be found randomly
- If PAR < RND < HMCR: a new combination is found by randomly selecting a value from the existing sets.
- If PAR > RND: adjust the obtained combination in option 2.

At the end of the selection, the objective function of the new combination is evaluated and compared against the worst combination in the existing sets. Once the required number of iterations is reached, the improvement process stops. Finally, the best combination with sets is identified and exported.

Code 8.7. Harmony search with two variables. Section 3: generating and comparing the new combination.

```
---------------
For iter = 1 To MaxIte-1
Randomize: rand = Rnd
For j = 1 To ND
            If rand >= HMCR Then 'Random Searching
                  x(j) = limit(j, 1) + (limit(j, 2)-
limit(j, 1)) * rand
            Else
                  d1 = Int(HMS * rand) + 1
                  x(j) = HM(d1, j)
                  If rand < PAR Then 'Pitch Adjusting
                        d1 = (limit(j, 2)-limit(j, 1))/
limit(j, 3)
                        x(j) = x(j) + d1 * (rand-PAR /
2)
                  End If
            End If
      Next j
sol = diffR2(x(1), x(2), feed, fraction, Conv, W)

'----find hmax and corresponding index hmax _ num
hmax _ n = 1: hmax = HM(1, 3)
For i = 2 To HMS
      If HM(i, 3) > hmax Then
            hmax _ n = i: hmax = HM(i, 3)
      End If
Next i
'----replacing hmax with the new solution if improved
```

```
If sol < hmax Then 'if new Sol is smaller than hmax,
replace hmax
    For j = 1 To ND
          HM(hmax _ n, j) = x(j)
    Next j
    HM(hmax _ n, 3) = sol
End If

Next iter

hmin _ n = 1: hmin = HM(1, 3)
For i = 2 To HMS
    If HM(i, 3) < hmin Then
          hmin _ n = i: hmin = HM(i, 3)
    End If
Next i
          out(3, 1) = "Number of iter": out(3,
2) = "x1": out(3, 3) = "x2": out(3, 4) = "Error"
          out(4, 1) = iter: out(4, 2) = HM(hmin _ n,
1): out(4, 3) = HM(hmin _ n, 2): out(4, 4) = HM(hmin _ n,
3)
harmony2 = out
End Function
```

In this code section, RANDOMIZE is used before calling the RND again. The value ("rand") is used for the three calculations in the three improving options. Technically, different random numbers can be generated and used for these calculations, which might improve the solution.

With limits at 3 and 10, the code can generate the answers as in Table 8.2. It should be noted that the answers will be different when the code is repeated. The two flow rates are close to the answer from the secant method and can be improved by increasing the number of iterations and other parameters (HMS, HMCR and PAR).

TABLE 8.2
Harmony Search Output

Min from the Initial Sets	Max from the Initial Sets	0	0
0.099774994	113.9915897	0	0
Number of iter	x1	x2	Error
10000	3.105824475	7.288284783	0.098218223

8.5   FURTHER EXAMPLES FOR SELF-PRACTICE

EXAMPLE 8.4.   REDUCING CALCULATIONS OF TEAR ANALYSIS.

In Example 8.2, each iteration must call functions difR1 and difR2 three times each. The calculations within these two functions are a sequence of mass balances for the modular operations of the process. The calculation of the two functions is essentially the same and thus offers an opportunity to reduce computational time. Rewrite the codes so each iteration only needs three mass balance calculations.

Answer

First, a new function is needed to replace both difR1 and defR2. The function will return two values simultaneously as two elements of an output array (2-dimensional). Consequently, the main function will call the corresponding elements in a Jacobian procedure. Therefore, the output array, even with only two elements, needs to be in a 2-dimensional form.

EXAMPLE 8.5.   Adjusting parameters of the harmony search.

In Example 8.3, the parameters are HMS = 100, HMCR = 0.9, PAR = 0.6, and MaxIte = 100000. Adjust HMS and MaxIte to improve the process.

Answer

By increasing the number of sets, the best solution is improved significantly. For instance, Table 8.3 shows that HMS = 500 and MaxIte = 100 can provide a better solution. However, in this case, the harmony search does not improve the best solution from the random set. Perhaps some of these problems can be solved by many random trials of guesses, similar to the search algorithm in Example 12.3.

TABLE 8.3
Harmony Search Output

Min from the Initial Sets	Max from the Initial Sets	0	0
0.006903335	126.048336	0	0
Number of iter	x1	x2	Error
100	3.161868572	9.779379845	0.006903335

---

9 Solving Ordinary Differential Equations

9.1 INTRODUCTION

An ordinary differential equation (ODE) is a differential equation containing one or more functions of one independent variable and the derivatives of those functions. The equations are used to describe many physical processes. For example, the equations play a central role in predicting the specie balances for chemical reactions. Consequently, solving ODEs is fundamental to reactor design and process flow sheeting. The order of the ODEs is defined as the highest order of any differential term. Any n^{th}-order ODE can be expressed in the system of n first-order ODEs.

Numerical methods for solving ordinary differential equations are techniques used to approximate the solution of an ODE. These methods involve approximating the derivative or differential equation with a mathematical expression that can be solved numerically. Many numerical schemes are available for solving ODEs, each with advantages and disadvantages.

9.2 ONE-STEP METHODS WITH A CONSTANT STEP SIZE

Assuming that we need to approximate the following ODE from x_0 to x_f:

$$\frac{dy}{dx} = f(x, y)$$

Eq. 9.1

where $y = y_0$ at $x = x_0$.

The one-step method will divide the domain, from $x = x_0$ to x_f, into m steps with a constant step size of h. The relationship between h and m is defined as:

$$h = (x_f - x_0)/m$$

Eq. 9.2

The method will then approximate the values of y at those steps sequentially. These one-step approximation methods are named the Runge–Kutta (RK) methods for differential equations. There are many variations, which are sub-classified according to the order of the scheme. For example, an nth-order RK method will require n evaluations within each step and has local and global errors on the order of h^{n+1} and h^n, respectively. There is only one first-order RK method, Euler's, which is the most straightforward numerical method for solving ODEs.

Mathematically, Euler's prediction is:

$$y(x+h) = y(x) + f(x,y)h$$

Eq. 9.3

This equation can be demonstrated in the following example.

EXAMPLE 9.1.   EULER'S METHOD FOR A SINGLE FIRST-ORDER ODE.

PROBLEM STATEMENT

Solve the equation by the Euler method:

$$dy/dx = \cos(x) \text{ from } x = 0 \text{ to } x = \pi$$

with an initial value $y = 0$ at $x = 0$.

Apply three different step sizes: $\pi/2$, $\pi/6$ and $\pi/12$. Then plot the results and compare them with the exact solution $y = \sin(x)$.

Solution

Using Euler's method, the solution can be found with simple VBA code (Code 9.1).

Code 9.1. Euler's method for a single first-order ODE.

```
Function RK_1(x0, y0, dx, xmax)
Dim matrix(), i, n

n = xmax/dx: ReDim matrix(n, 1)
matrix(0, 0) = x0: matrix(0, 1) = y0:

For i = 1 To n
matrix(i, 0) = matrix(i-1, 0) + dx
matrix(i, 1) = matrix(i-1, 1) + ff(matrix(i-1, 0),
matrix(i-1, 1)) * dx
Next i
RK_1 = matrix
End Function

Function ff(x, y)
ff = Cos(x)
End Function
```

It should be noted that the function **ff** in this example depends on x only, so y can be removed from the argument of function **ff**. One of the apparent problems with Euler's method is balancing the accuracy and computational time. This can be demonstrated by the three solutions in Figure 9.1.

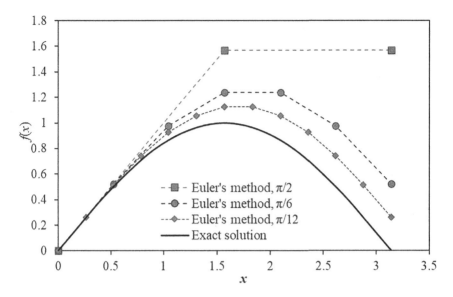

FIGURE 9.1   Influence of step size on the accuracy of Euler's method.

Figure 9.1 shows that solution accuracy increases with the reduced step size, h. However, smaller step sizes will increase the number of iterations and, thus, computational time. The curves also demonstrate two types of errors in the method: the local truncation error and the accumulative truncation error. For Euler's method, the local error is on the order of h^2, and the global error is on the order of h (Chapra and Canale 2010). For many problems, the required step size might be too small and result in numerous iterations. Consequently, there are other RK methods in the literature.

9.2.1   SECOND-ORDER METHODS

In any other order higher than 1, the RK method has many variations. For example, the second-order Runge–Kutta method has three well-known variations: Heun's, improved polygon and Ralson's schemes (Chapra and Canale 2010). These variations have similar accuracy and computation complexity. Here equations and VBA code for the improved polygon are introduced.

Mathematically, the improved polygon method predicts:

$$y_{i+1} = y_i + f\left(x_{i+0.5}, y_{i+0.5}\right)h \qquad \text{Eq. 9.4}$$

where $y_{i+0.5}$ is the slope of the midpoint between x_i and x_{i+1} and is given by:

$$y_{i+0.5} = y_i + f\left(x_i, y_i\right)h/2 \qquad \text{Eq. 9.5}$$

VBA code for the improved polygon method is given in Code 9.2.

Code 9.2. Improved polygon method for a single first-order ODE.

```
Function RK_2(x0, y0, dx, xmax) 'Improved polygon
Dim matrix(), i, n, x_m, y_m

n = xmax/dx: ReDim matrix(n, 1)
matrix(0, 0) = x0: matrix(0, 1) = y0:

For i = 1 To n
matrix(i, 0) = matrix(i-1, 0) + dx
x_m = matrix(i-1, 0) + dx / 2
y_m = matrix(i-1, 1) + ff(matrix(i-1, 0),
matrix(i-1, 1)) * dx / 2
matrix(i, 1) = matrix(i-1, 1) + ff(x_m, y_m) * dx
Next i
RK_2 = matrix
End Function
```

9.2.2 FOURTH-ORDER METHODS

There are many variations of Runge–Kutta methods in third, fourth and higher orders. Yet there is one outstanding fourth-order method. This method is sometimes referred to as the classical fourth-order RK method. The scheme is given as:

$$y_{i+1} = y_i + \left[\frac{1}{6}\left(k_1 + 2k_2 + 2k_3 + k_4\right)\right]h \qquad \text{Eq. 9.6}$$

where

$$k_1 = f\left(x_i, y_i\right) \qquad \text{Eq. 9.7a}$$

$$k_2 = f\left(x_i + \frac{1}{2}h, y_i + \frac{1}{2}k_1 h\right) \qquad \text{Eq. 9.7b}$$

$$k_3 = f\left(x_i + \frac{1}{2}h, y_i + \frac{1}{2}k_2 h\right) \qquad \text{Eq. 9.7c}$$

$$k_4 = f\left(x_i + h, y_i + k_3 h\right) \qquad \text{Eq. 9.7d}$$

VBA code for the method is given in Code 9.3. The code and VBA functions in Codes 9.1 and 9.2 compare the three methods, as demonstrated in Figure 9.2.

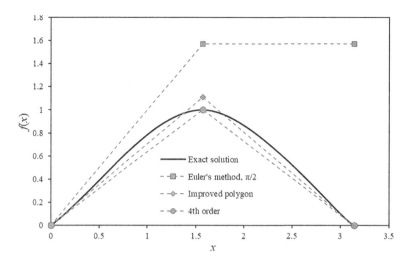

FIGURE 9.2 Comparison between Euler's, improved polygon and fourth-order Runge–Kutta methods.

Code 9.3. The classical fourth-order RK method.

```
Function RK _ 4(x0, y0, h, Xmax)
Dim m(), i, n
Dim k1, k2, k3, k4
n = Xmax/h: ReDim m(n, 1)
m(0, 0) = x0: m(0, 1) = y0:

For i = 1 To n
m(i, 0) = m(i-1, 0) + h
k1 = ff(m(i-1, 0), m(i-1, 1))
k2 = ff(m(i-1, 0) + h / 2, m(i-1, 1) + k1 * h / 2)
k3 = ff(m(i-1, 0) + h / 2, m(i-1, 1) + k2 * h / 2)
k4 = ff(m(i-1, 0) + h, m(i-1, 1) + k3 * h)
m(i, 1) = m(i-1, 1) + (k1 + 2 * k2 + 2 * k3 + k4) * h / 6
Next i
RK _ 4 = m
End Function
```

9.3 ONE-STEP METHODS WITH AN ADAPTIVE STEP SIZE

The previous section shows that increasing the orders of Runge–Kutta methods can reduce the numerical steps without compromising accuracy. However, these methods still require a constant step size. In these instances, the number of steps and the final step are selected regardless of the behaviour function. However, the actual local error depends on the local behaviour. For instance, if the function is almost linear, a moderate step size is adequate. However, a smaller step size might be required if the equations change dramatically.

A general equation might have sections with dramatic and gradual changes. Consequently, the coder must balance step size (and the number of steps) and accuracy. One way to achieve this is by using an adaptive step size. In this approach, the step size of each increment is determined by estimating the local absolute error.

If only one RK method is used, the absolute error is unknown. However, combining two RK methods with different orders can control the error. In theory, any two RK methods can be selected. However, there are certain combinations in which the evaluations are *nested* to reduce the number of function evaluations. One of the well-known combinations is the Runge–Kutta Fehlberg method, which combines a fourth- and fifth-order Runge–Kutta method.

On the other hand, the easiest way is combining a first-order and second-order method. This section uses the combination to demonstrate the adaptive step size. Within each iteration, the method starts with a guessed step size, $h_{present}$. Consequently, the absolute difference between the full-step and the two half-step values is the error estimate, err_c. By comparing with the user-specified tolerance, err_d, the step size is adjusted (Press and Teukolsky 1992):

$$h_{new} = 0.95\, h_{present} \left[\frac{err_d}{err_c} \right]^{\alpha} \qquad \text{Eq. 9.8}$$

In Eq. 9.8, α is the adjustment factor, which equals 0.2 when $err_c > err_d$, and 0.25 when $err_c < err_d$. Using the formula, the step size is adjusted until the desired accuracy is given by

$$err_d = err_a\, y(x + h) \qquad \text{Eq. 9.9}$$

where err_a is the required relative accuracy.

EXAMPLE 9.2.   RK METHODS WITH AN ADAPTIVE STEP SIZE.

PROBLEM STATEMENT

Compare different Euler's methods for the following ODE:

$$y' = 15 e^{-10(x-2)^2} - 0.8y$$

from $x = 0$ to $x = 4$ with an initial value $y = 0.5$ at $x = 0$.
Part A. Using Euler's method with a step size of 0.4.
Part B. Using the classical fourth-order Runge Kutta method with a step of 0.5.
Part C. Using an adaptive step size from Euler's and improved polygon methods.
Plot the results and compare the results.

Solution

The Euler method in Code 9.1 can be modified to combine with function *fff* in Code 9.4. The fourth-order Runge–Kutta and adaptive methods are presented in Code 9.4.

Code 9.4. Adaptive step size with Euler and second-order method.

```
Option Base 0
Function fff(x, y)
fff = 15 * Exp(-(x-2) ^ 2 / 0.1)-0.75 * y
End Function
Function RK _ 42(x0, y0, h, Xmax)
Dim m(), i, n, k1, k2, k3, k4

n = Xmax/h: ReDim m(n, 1)
m(0, 0)  = x0: m(0, 1)  = y0:

For i = 1 To n
m(i, 0)  = m(i-1, 0)  + h
k1 = fff(m(i-1, 0), m(i-1, 1))
k2 = fff(m(i-1, 0)  + h / 2, m(i-1, 1)  + k1 * h / 2)
k3 = fff(m(i-1, 0)  + h / 2, m(i-1, 1)  + k2 * h / 2)
k4 = fff(m(i-1, 0)  + h, m(i-1, 1)  + k3 * h)
m(i, 1)  = m(i-1, 1)  + (k1 + 2 * k2 + 2 * k3 + k4) * h / 6
Next i

RK _ 42 = m
End Function


Function adaptive _ E _ 2(x, y, h, errA)
Dim del, y1, y05, x05, output(3), k, ErrD, hnew
ErrD = 1: del = ErrD * 2: hnew = h: k = 0

While del > ErrD And k < 200
h = hnew: k = k + 1
y1 = y + fff(x, y) * h
y05 = y + fff(x, y) * h / 2
del = Abs(fff(x + h / 2, y05) * h / 2 + y05-y1)
ErrD = errA * Abs(y1)

If Abs(del) > ErrD Then
hnew = 0.95 * h * Abs(ErrD/del) ^ 0.25
Else
hnew = 0.95 * h * Abs(ErrD/del) ^ 0.2
End If
Wend

output(0)  = h: output(1)  = x + h: output(2)  = y1:
output(3)  = k
adaptive _ E _ 2 = output
End Function
```

TABLE 9.1

Adaptive Step Size

A	B	C	D
h	*x*	*y*	*k*
0.5	0	0.5	
0.23497	0.23497	0.411886106	4
0.23497	0.469941	0.339300329	4
0.23497	0.704911	0.279506296	4
0.234811	0.939721	0.230283213	4
0.219514	1.159236	0.192413513	4
0.119156	1.278392	0.176739477	4
0.064968	1.34336	0.173465187	5
0.048407	1.391767	0.17690453	5
0.03784	1.429607	0.185924458	6
0.033484	1.463091	0.200662556	6
0.030896	1.493988	0.221957736	6
0.027883	1.521871	0.249633028	7
0.027085	1.548956	0.285865638	7
0.026812	1.575768	0.33270589	7
0.02698	1.602748	0.392888656	7
0.027554	1.630302	0.470063829	7
0.028543	1.658845	0.569148304	7
0.030004	1.68885	0.696886125	7

Please note that the **adaptive_E** function in Code 9.4 only does one step. To get the whole function from 0 to 4, the function in Code 9.4 is called continuously in a table form. A section of the table is shown in Table 9.1. The VBA function also gives the number of adjustments, k, at each step. The predetermined error plays a clear role in this value. In Table 9.1, the starting step, h, is fixed at 0.5, and the required relative error, $ErrA$, is 1%. The code is designed so that this input can be changed easily.

The adaptive method is plotted against Euler's, fourth-order RK and analytical solutions in Figure 9.3.

At first glance, the adaptive step method requires many iterations, with some improved accuracy over Euler's method. However, the combined first- and second-order methods have errors on the order of h^3, which is still larger than a fourth-order Runge–Kutta method. Hence, the adaptive method is less accurate than the fourth-order RK, as shown in Figure 9.3.

On the other hand, the adaptive method has an advantage when the whole function is needed. For instance, when function behaviour is used to integrate a curve under the function, the accuracy along the function is more important than the final value at the upper limit. This is demonstrated in the plug-flow reactor design in Chapter 10.

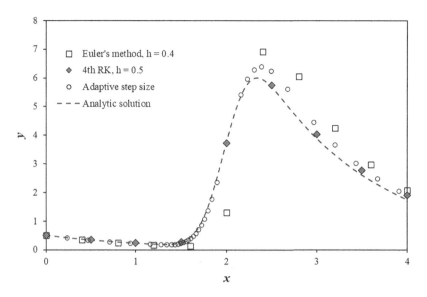

FIGURE 9.3   RK method with an adaptive step size.

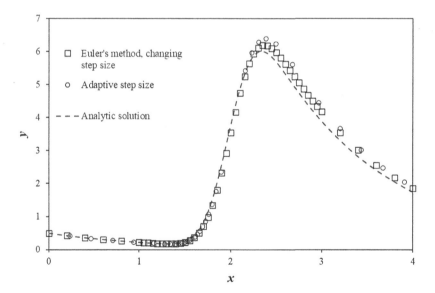

FIGURE 9.4   Euler's method with a changing step size.

Finally, the calculation can be optimised by manually selecting the step size. For example, after applying Euler's method with a constant step size of 0.4, dramatic changes are observed in the region between 1 and 3. Consequently, the whole region between 0 and 4 can be divided into three sections: (0,1), (1,3) and (3,4). The step sizes in these three regions are selected at 0.2, 0.05 and 0.2, respectively. The resulting prediction with Euler's method is demonstrated in Figure 9.4. The method has a better prediction than the adaptive method while having a very simple calculation. The approach requires a visual justification of the important regions. Such an approach is very helpful for a system of ODEs, such as reactor modelling in Chapter 10.

9.4   JUSTIFICATION OF THE SELECTED STEP SIZE

In the previous sections, it can be seen that the step size plays a critical role in determining the accuracy of the numerical methods. It is difficult to justify the step size without knowing the actual function (as in Figures 9.3 and 9.4). Generally, a smaller step size is better. Yet there is no clear definition of a *sufficiently* good step size. Selecting an unnecessarily small step size would significantly increase the number of steps.

One simple approach is applying several different step sizes until no *significant* changes occur. Here, the approach is applied to the previous example. Assume that Euler's method with three constant sizes is selected. For the two regions, $(0,1)$ and $(3,4)$, h_0 is used. Between 1 and 3, the step size is reduced four times, $h_0/4$. Subsequently, the method is applied for four values of h_0: 0.4, 0.2, 0.1 and 0.05 (Figure 9.5). In Figure 9.5, it can be seen that the last two curves are almost identical. Consequently, an h_0 value of 0.1 should be sufficiently adequate for the problem.

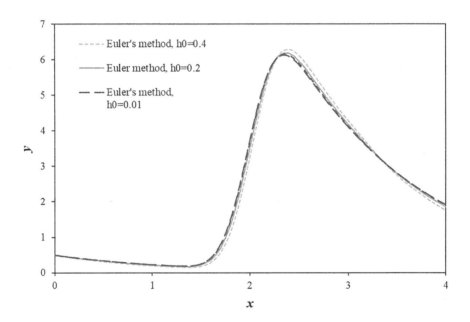

FIGURE 9.5   Euler's method with four different step sizes.

9.5   CONCLUDING REMARKS

While there are many different numerical methods for solving ODEs, all methods have advantages and disadvantages. The users need to balance between accuracy, computational time and complication of the coding. The third factor will be demonstrated in the next chapter, in which the solutions are combined with other numerical methods to solve engineering problems. Euler's method with a reasonable step size should be used. If the method is inadequate, the second preferable option should be the fourth-order RK (Code 9.3). Finally, if the function has a significant variation in a certain region, a manual step size variation should be employed.

10 Applications of ODEs

10.1 INTRODUCTION

Numerical methods for solving ODEs are essential for designing and analysing engineering systems. Numerical methods are used to simulate these models and make predictions about the behaviour of the physical system. With a routine solution of a system of ODEs, the VBA code can be modified to solve complicated problems, such as optimising a reactor. In particular, combining other methods, such as bisection or binary search, can provide a powerful tool to solve these physical processes.

10.2 ADVANCED EXAMPLE: OPTIMISATION OF A SEMI-BATCH REACTOR

A semi-batch process, also referred to as a semi-continuous process, is an operational mode used in chemical and process engineering. It combines elements of both batch and continuous processing. In a batch process, all the raw materials are added at the start, the reactions occur and the product is removed at the end. In a continuous process, raw materials are constantly added, and products are constantly removed.

A semi-batch process is a middle ground between these two extremes. In this process, one or more reactants are fed into the reaction mixture over time. This might be done to control the reaction rate, manage heat release from an exothermic reaction or optimise the yield or selectivity of a reaction. For example, if a reaction is highly exothermic (i.e. it releases a lot of heat), adding all the reactants at once could cause the reaction to be uncontrolled, potentially leading to a dangerous situation (Fogler 2006). By adding one reactant slowly, the heat release can be managed safely. Similarly, if a reaction produces an undesired by-product, slowly adding one reactant can help minimise its production. Modelling a semi-batch process often requires simultaneously solving a system of differential and ordinary equations.

EXAMPLE 10.1. SEMI-BATCH REACTOR.

PROBLEM STATEMENT

Consider the series of two reactions, adapted from Example 9.7 (Fogler 2006):

$2A \xrightarrow{k_1} B$ heat of reaction $\Delta H_1 = -6500$ cal/mol A, rate of reaction k_{1a}

$B \xrightarrow{k_2} 3C$ heat of reaction $\Delta H_2 = 8000$ cal/mol B, rate of reaction, k_{2b}

$C_{PA} = 30$ cal/mol K, $C_{PB} = 60$ cal/mol K, $C_{PC} = 20$ cal/mol K, $C_{PCatalyst} = 35$ cal/mol K

    DOI: 10.1201/9781003364511-10

The feed is added continuously at a volumetric flow rate of 240 L/h and temperature of 305 K. The feed contains A at a concentration of 4 mol/L. The reactor has a heat exchanger with UA = 35000 cal/h K and an exchange (air) temperature of 298 K. Initially, there is 100 dm^3 solution in the reactor, with 1 mol/dm^3 of A and 1 mol/L of the catalyst.

Part A. Calculate and plot the chemical concentrations and temperature as a function of time from zero to 1.5 hours.

Part B. Calculate the maximum concentration of B and the corresponding reaction time.

Solution

Part A. The problem can be solved using a system of four ODEs and several equations. The four ODEs account for the changes of three chemicals and temperature. Heat balance equations describe the transient heat of the two reactions and the heat input. First, the values of parameters are input into an Excel spreadsheet (Table 10.1).

TABLE 10.1
Parameters of Reactions, Flow Rate, Temperature and Initial Conditions

	A	B	C
1	**Reactor Variables**		Unit
2	*Cao*	4	mol/dm^3
3	*To*	305	K
4	*vo*—flowrate	240	dm^3/h
4	*Ta*—temp of air	298	K
5	*V_in*	100	dm3
6	*R*	1.987	cal·K^{-1}·mol^{-1}
7	*E1*	9500	cal/mol
8	*E2*	7000	cal/mol
9	*Tref1*	320	K
10	*Tref2*	300	K
11	k_{1a}	1.25	h^{-1}
12	k_{2b}	0.08	h^{-1}
13	**Energy variables**		
14	*delHra*	−6500	cal/mol A
15	*delHrb*	8000	cal/mol B
16	*UA*	35000	cal/hK
17	*Cpa*	30	cal/molK
18	*Cpb*	60	cal/molK
19	*Cpc*	20	cal/molK
20	*Cpcatalyst*	35	cal/molK
21	*N_cat*	100	mol

While these parameters can be included in the VBA function as arguments, such an arrangement will make the function more complicated and prone to error when calling the macro. In this instance, the values of parameters are assigned to the VBA variable using the Range object. The positions of the cells need to be fixed.

Code 10.1. ODEs to calculate the concentration and temperature profiles of a semi-batch reactor.

```
Option Base 0
Function dabcTdt(dt, tmax)
Dim M()
Dim i, n, rA, rB, rC, dadt, dbdt, dcdt, dTdt, r1, r2,
CA0, CB0, CC0, T0, Ta, R
Dim E1A, E2B, T1, T2, k1a, k2b, delH1, delH2, UA
Dim CpA, CpB, CpC, CpCa, v0, V_in, CAf, H_ex,
H_A, H_r
'----------input
CAf = Range("B2").Value
T0 = Range("B3").Value
v0 = Range("B4").Value
Ta = Range("B5").Value
V _ in = Range("B6").Value
R = Range("B7").Value
E1A = Range("B8").Value: E2B = Range("B9").Value:
T1 = Range("B10").Value: T2 = Range("B11").Value:
k1a = Range("B12").Value: k2b = Range("B13").Value:
delH1 = Range("B15").Value: delH2 = Range("B16").Value:
UA = Range("B17").Value:
CpA = Range("B18").Value: CpB = Range("B19").Value:
CpC = Range("B20").Value: CpCa = Range("B21").Value:
'--------------
n = tmax/dt: ReDim M(0 To n, 0 To 5)
M(0, 1) = V _ in: M(0, 2) = 1: M(0, 3) = 0: M(0, 4) = 0:
M(0, 5) = 290

For i = 1 To n
M(i, 0) = M(i-1, 0) + dt: M(i, 1) = M(i-1, 1) + dt * v0
r1 = -k1a * Exp(E1A/R * (1 / T1-1 / M(i-1, 5))) *
M(i-1, 2)
r2 = -k2b * Exp(E2B/R * (1 / T2-1 / M(i-1, 5))) * M(i-1, 3)

dadt = r1 + (CAf-M(i-1, 2)) * v0 / M(i-1, 1): M(i,
2) = M(i-1, 2) + dadt * dt
dbdt = -r1 / 2 + r2-M(i-1, 3) * v0 / M(i-1, 1): M(i,
3) = M(i-1, 3) + dbdt * dt
```

```
dcdt = -3 * r2-M(i-1, 4) * v0 / M(i-1, 1): M(i,
4) = M(i-1, 4) + dcdt * dt

H _ ex = UA * (298-M(i-1, 5)) 'heat loss to heat exchanger
H _ A = CAf * v0 * CpA * (M(i-1, 5)-T0) 'heat of inflow
of pure A
H _ r = (delH1 * r1 + delH2 * r2) * M(i-1, 1)
'-need to calculate two terms for energy balance
dTdt = (H _ ex-H _ A + H _ r)/((CpA * M(i-1, 2) + CpB *
M(i-1, 3) + CpC * M(i-1, 4)) * M(i, 1) + 100 * CpCa)
M(i, 5) = M(i-1, 5) + dt * dTdt
Next i
dabcTdt = M
End Function
```

Part B. From the graph in Figure 10.1, the concentration of B has a maximum of around 0.6 h. To determine this point precisely, we can combine the ODE code with region elimination code.

First, the function in Code 10.1 is revised so that the output only contains concentration B at time $tmax$. The new function can be named **dablTdt_2** (Code 10.2). The function **dabcTdt_2** is essentially the same as in Code 10.1, except for output $(M(n,3)$ instead of $M)$. Consequently, the function is a subfunction within a golden section search. The final answers are $C_{B\_max} = 1.191$ and $t_m = 0.6438$ h.

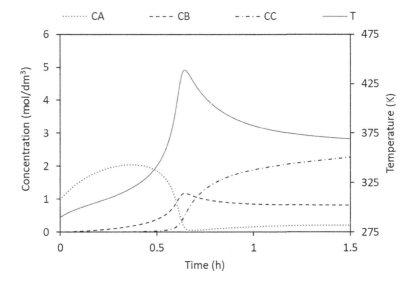

FIGURE 10.1   Concentrations and temperature of a semi-batch reactor.

Code 10.2. Combination of golden section search and ODEs to optimise a semi-batch reactor.

```
Option Base 0
Function dabcTdt _ 2(dt, tmax)
'----------input
n = Round(tmax/dt, 0): ReDim M(0 To n, 0 To 5)

'------------same as in Code 10.1
dabcTdt _ 2 = M(n, 3)
End Function


Function golden _ section(t _ l, t _ h, imax, delt)
Dim Gr, d, D1, D2, f1, f2, i, output(2)
Gr = (5 ^ 0.5-1) / 2:
d = t _ h-t _ l: i = 0
d = d * Gr: D2 = t _ l + d: D1 = t _ h-d:
f1 = dabcTdt _ 2(delt, D1): f2 = dabcTdt _ 2(delt, D2)

While d > delt / 2 And i < imax
d = d * Gr
If f1 < f2 Then
    t _ l = D1: D1 = D2: D2 = t _ l + d
    f1 = f2: f2 = dabcTdt _ 2(delt, D2)
Else
    t _ h = D2: D2 = D1: D1 = t _ h-d
    f2 = f1: f1 = dabcTdt _ 2(delt, D1)
End If
i = i + 1
Wend

output(0) = (t _ l + t _ h) / 2: output(1) =
dabcTdt _ 2(delt, output(0))

golden _ section = output
End Function
```

It should be noted that the "**For Next**" loop in VBA will stop at the rounded number (for example, $n = 12.6$, the loop stops at 12). So, if the step size of ODE is larger than the tolerance of the golden section, the two internal points D_1 and D_2 can be the same and may lead to a wrong maximum. The second notation is that the **ReDim** of $M(0$ to $n)$ gives the dimension of the nearest integer to n (for example, if $n = 12.6$, M has 13 columns).

Similarly, the index is rounded to the nearest integer, $M(12.6,3) = M(13,3)$. Consequently, using **round** in Code 10.2 is needed to avoid the error. Since n is rounded, the step in the golden section needs to be larger than delt/2 (to differentiate the values of $f1$ and $f2$).

10.3   ADVANCED EXAMPLE: BOUNDARY VALUE ODEs

A boundary value problem (BVP) for ordinary differential equations is a type of problem where the solution of the ODE is sought subject to specified boundary conditions. In contrast to initial value problems, where the values of the solution and its derivatives are specified at a single point, in BVPs, the values of the solution are specified at different points. BVPs arise in many areas of science and engineering, including mechanics, physics, chemistry and biology.

The two-point BVP is the most common boundary value problem for ODEs, where the solution is required to satisfy specified conditions at different points. To solve a BVP, we can combine the numerical methods for initial value problems with a root-of-equation method. This approach requires the (computational) solution of an initial value problem. The method assumes the initial values of the unknown variables and subsequently adjusts the value to satisfy the condition at the other boundary. The adjustment can be done by using the bisection method. The procedure is often referred to as a "shooting method".

EXAMPLE 10.2.   SHOOTING METHOD FOR 1-D HEAT DISTRIBUTION.

PROBLEM STATEMENT.

Considering a rod 10 cm long, the temperatures of two ends are maintained at 40 and 200°C. The rod is exposed to air at a constant temperature, $Ta = 20°C$. At a steady state, the heat transfer is governed by the equation:

$$\frac{d^2T}{dx^2} = h(T_a - T)$$ Eq. 10.1

where h is the radiative heat loss parameter (cm^{-1}).
Calculate the temperature profile along the rod.

Solutions

The previous second ODE can be expressed as a system of two first ODEs:

$$\frac{dT}{dx} = z$$ Eq. 10.2

$$\frac{dz}{dx} = h(T_a - T)$$ Eq. 10.3

In this problem, we will use two different VBA codes. First, code is developed to calculate the temperature profile for a given set of initial values (Code 10.3). The code treats the problem as a system of two first ODEs. The code can provide the temperature profile at the given heat transfer dz/dx ($x = 0$).

Code 10.3. Solving the temperature contribution along a long rod for a given initial conditions.

```
Option Base 0
Function temp(t0, z0, Ta, h, dx, xmax)
Dim output(), i, n, dxdt, dzdt
n = xmax/dx: ReDim output(n, 2)
output(0, 0) = 0: output(0, 1) = t0: output(0, 2) = z0

For i = 1 To n
output(i, 0) = output(i-1, 0) + dx
output(i, 1) = output(i-1, 1) + output(i-1, 2) * dx
output(i, 2) = output(i-1, 2) + h * (output(i-1, 1)-Ta) * dx
Next i
temp = output
End Function
```

The second code (Code 10.4) uses using bisection method to work out the required slope ($z0$ or dT/dt at $x = 0$) so that the temperature at the other end equals $200°C$.

Code 10.4. Solving the temperature contribution along a long rod by a bisection method.

```
Function f(t0, z0, Ta, h, dx, xmax, Te)
Dim m, n
n = xmax/dx
m = temp(t0, z0, Ta, h, dx, xmax)
f = Te-m(n, 1)
End Function

Function BisectionX(Lo, Up, tolt, t0, Ta, h, dx, xmax, Te)
Dim Mi, c1, c3, i, k
k = Abs(Up-Lo)/tolt
c1 = f(t0, Lo, Ta, h, dx, xmax, Te)
For i = 1 To k
Mi = (Lo + Up) / 2: c3 = f(t0, Mi, Ta, h, dx, xmax, Te)
If c1 * c3 < 0 Then
    Up = Mi
Else
    Lo = Mi: c1 = c3
End If
Next i
BisectionX = Mi
End Function
```

The function f in Code 10.4 uses the *temp* function in 10.3. The output of f is the difference between the required temperature (200°C) and the predicted value at $x = 10$. Using the function **bisectionX**, one gets a value of 13.413 so that $f = 0$. Using this value as input $z0$, Code 10.3 will give the answer in Table 10.2 and Figure 10.2.

TABLE 10.2

Temperature Distribution along a Conductive Rod

	A	B	C
1	Distance (m)	Temperature (°C)	Gradient (°C/m)
2	0	40	13.41394
3	1	53.41394	13.61394
4	2	67.02788	13.94808
4	3	80.97596	14.41836
5	4	95.39433	15.02812
6	5	110.4224	15.78206
7	6	126.2045	16.68629
8	7	142.8908	17.74833
9	8	160.6391	18.97724
10	9	179.6164	20.38363
11	10	200	21.9798

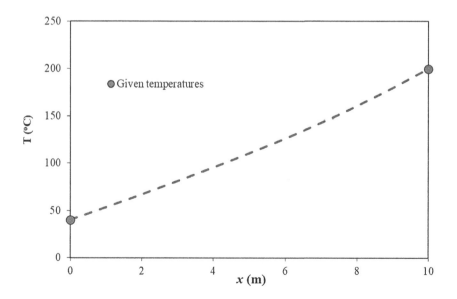

FIGURE 10.2   Temperature distribution along a conductive rod.

The exact solution for the temperature is given (Chapra and Canale 2010).

$$T(x) = 73.4523\ e^{0.1x} - 53.4523e^{-0.1x} + 20 \qquad\qquad \text{Eq. 10.4}$$

The user can compare the numerical solution to this exact solution.

10.4 FURTHER EXAMPLES FOR SELF-PRACTICE

EXAMPLE 10.3. TEST THE REDIM OF AN ARRAY.

Reconsidering Code 10.2, if the value of n in function **dabcTdt_2** is calculated as *tmax*/delt, what would happen to the function **golden_section**?

Answer

The program may give an error depending on the value of n, which is decided by the selected value of *dt*. For example, if $n = 12.51$, the array will have 13 rows. Hence, $M(n,5)$ becomes $M(13,5)$. However, the **For Next** loop stops at 12.

EXAMPLE 10.4. OPTIMISING A FIRST-ORDER ADIABATIC BATCH REACTOR, ADAPTED FROM EXAMPLE 9.1 (FOGLER 2006).

For a first-order adiabatic batch reactor, the rate of change of conversion is given by:

$$\frac{dX}{dt} = k(1 - X)$$

where k is the reaction rate constant, which is dependent on the temperature

$$k = (2.73 \times 10^{-4}) \exp\left[\frac{32,400}{1.987}\left(\frac{1}{535} - \frac{1}{T}\right)\right]$$

where reactor temperature, T, is the function of conversion

$$T = 515 + 90.1\,X$$

Calculate the residence time and reactor temperature for the conversion of $X = 52\%$.

Hints

VBA code is needed to solve the system of three equations. Although the first equation is an ordinary differential equation, the other two are explicit equations (these two will be substituted into the ODE). Hence, code for a single ODE can be used. Furthermore, the code can produce the T and X for plotting (Figure 10.3).

The code can be combined with a bisection method to find time and temperature when X is 52 %. A bisection code can be used. Hence, the bisection code must evaluate X at different values of t. It should be noted that each of these evaluations uses the same calculation from the beginning, $t = 0$. Consequently, the scanning method can be used to reduce the calculations.

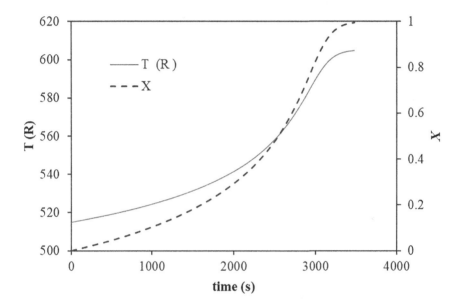

FIGURE 10.3   Temperature and conversion as functions of time.

11 Integration

11.1 INTRODUCTION

Numerical integration is the process of approximating the definite integral of a function using numerical methods. It is a widely used technique in mathematics, science and engineering, where analytical solutions may be difficult or impossible.

For a positive function, $f(x) > 0$, the integration of $f(x)$ from a to b equals the area under curve $f(x)$ between a and b. This chapter applies the integration to find the area under a curve. The calculation can be applied to other integration problems.

The most common numerical integration methods are based on approximating the area under the function's curve using geometric shapes, such as rectangles, trapezoids and Simpson's rules. The choice of method depends on the smoothness and complexity of the function being integrated and the desired level of accuracy.

Amongst these three, the Simpson rule (or Simpson's 1/3 rule) requires equal step size. The square method is suitable if the curve is relatively flat. The trapezoid method is suitable for steep but mostly linear curves. The equation for the trapezoid method is:

$$A = \frac{h}{2}\left(f(x_1) + f(x_2)\right)$$

Eq. 11.1

where h is the step size (distance between x_1 and x_2).

Alternatively, the value of the function at the midpoint can be used:

$$A = hf((x_1 + x_2)/2)$$

Eq. 11.2

Eq. 11.2 has an advantage over Eq. 11.1; it requires only one evaluation. The formula is more accurate than the simple rectangular method, which only requires $f(x1)$.

Simpson's rule involves approximating the function by a quadratic polynomial and integrating this polynomial over two neighbouring elements of the integration. In addition, Simpson's rule relies on quadratic interpolation over three points of the curve, thus requiring an even number of equal-distanced elements.

The area of two elements is given by:

$$A = \frac{h}{6}\left(f(x_1) + 4f(x_1 + h/2) + f\left(x_1 + h\right)\right)$$

Eq. 11.3

The equation is applied to the entire range between a and b.

In addition to the standard Simpson's rule, there is Simpson's 3/8 rule. The integration is based on a cubic interpolation over three elements in this case. The formula for the 3/8 rule is:

$$A = \frac{h}{8}\left(f(x_1) + 3f(x_1 + h/3) + 3f(x_1 + 2h/3) + f\left(x_1 + h\right)\right)$$

Eq. 11.4

DOI: 10.1201/9781003364511-11

11.2   AREA UNDER A CURVE FROM DISCRETE DATA POINTS

The integration can be obtained when the function is not given an explicit form. Such data are conveniently stored in an Excel spreadsheet.

EXAMPLE 11.1.   CALCULATE THE ACCUMULATIVE FLOW RATE.

PROBLEM STATEMENT

The available flow rate data is given in Table 11.1. Calculate the accumulative volume over the period.

TABLE 11.1
Volumetric Flowrate Data

	A	B
	A	*B*
1	Time (min)	Flow (L/min)
2	0	3.5
3	1.8	5.1
4	3.5	5.6
5	6	5.8
6	7	5.9
7	8.2	6.2
8	9.1	6.3
9	10.5	6.1
10	11.6	5.5
11	12	5

Solution

Since the recorded intervals are irregular, we cannot use Simpson's method. Hence, the trapezoid method (Eq. 11.1) can be used.

Code 11.1. Integration of area under a curve from discrete points.

```
Function Area(A)
Dim n, i
n = A.Rows.Count
For i = 2 To n
Area = Area + (-A(i-1, 1) + A(i, 1)) * (A(i-1, 2) +
A(i, 2)) / 2
Next i
End Function
```

Applying the code to Table 11.1, one can get a final answer of 7.74 L.

11.3　ADVANCED EXAMPLE: PLUG FLOW REACTOR PERFORMANCE

A plug flow reactor (PFR), or a tubular reactor, is a chemical reactor in which reactants are introduced at one end of a tube, and product(s) are collected at the other end of the tube. The tube typically has a uniform cross-sectional area. In addition, it is well mixed in the axial direction, meaning that the composition of the reactants and products does not vary significantly along the length of the tube.

In a PFR, the concentration of reactants and products at any point along the length of the tube is determined by the rate of the chemical reaction and the rate of mass transfer into and out of the tube. A plug flow reactor can be modelled by differential equations that describe the concentration profile of the species in the reactor as a function of time and axial position. Such systems can be solved using the methods in Chapters 9 and 10.

For many gas-phase reactions, however, the reactions can generate significant changes in gas volume. Therefore, the differential methods can be complicated for such varying-volume or varying-density systems. Alternatively, one can model this process by using the concept of space-time and the fractional change of volume (Levenspiel 1999).

EXAMPLE 11.2.　MODELLING A REACTOR CONVERSION FROM SPACE-TIME.

PROBLEM STATEMENT.

This problem is adapted from problem 5.4 (Levenspiel 1999).
Consider a homogenous gas reaction $A \rightarrow 3R$. The reaction has a rate at 215°C is:

$$-r_A = 10^{-2} C_A^{0.5} \text{ mol/L s}$$

The feed contains 50% A and 50% inert gas with $C_{AO} = 0.0625$ mol/L.
Part A. Determine the space-time needed for 80% conversion.
Part B. If the flow rate is increased so that space-time equals 35 s, what is the conversion of the reactor?

Solution

Part A. Space-time is a mathematical concept for flow reactors and is defined as the ratio between the feed and reactor volume. For example, a space-time of 30 seconds means the reactor treats one reactor volume of feed every 30 s.

For a plug flow reactor in this question, the space-time, τ, is related to the reaction rate and concentration of feed by:

$$\tau = \frac{C_{Ao}^{0.5}}{10^{-2}} \int_0^{X_{Af}} \left(\frac{1 + \varepsilon_A X_A}{1 - X_A} \right)^{0.5} dX_A \qquad \text{Eq. 11.5}$$

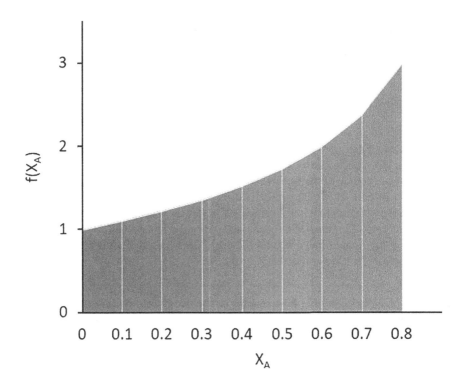

FIGURE 11.1   Plug flow reactor performance curve.

where X_A is the conversion and ε_A is the fractional change in the volume of the system. Since the feed contains 50% inert and the reaction triples the gaseous volume, ε_A equals 1. Consequently, Eq. 11.5 becomes:

$$\tau = \frac{C_{Ao}^{0.5}}{10^{-2}} \int_0^{X_A} \left(\frac{1+X_A}{1-X_A} \right)^{0.5} dX_A \qquad \text{Eq. 11.6}$$

The function inside the integral can be evaluated and plotted for X_A between 0 and 100%, as in Figure 11.1. Consequently, the integral in Eq. 11.5 is represented by the shaded area. The integral can be obtained by VBA code (Code 11.2).

Code 11.2. Space-time of a plug flow reactor by a trapezoid method.

```
Function trapez(x1, x2, n)
Dim i, dx
dx = (x2-x1)/n
For i = 0 To n-1
trapez = trapez + fXa(x1 + (i + 0.5) * dx) * dx
```

```
Next i
End Function
Function fXa(Xa)
fXa = ((1 + Xa)/(1-Xa)) ^ 0.5
End Function
```

In the code, fXa is defined in a separate VBA function. Using the code with $n = 8$ will return the shaded area as 1.324. Applying Eq. 11.5, $\tau = 33.1$ s. Increasing n will increase the accuracy of the method.

Part B. In part A, the integration can be obtained numerically for a given X_a. However, the question requires a value of X_a so that $\tau = 35$s. To get the solution, two macros are used. The first VBA code is integration, as in Code 11.2. The second one is applying the Regula Falsi method (Code 11.3). Using these VBA functions, the conversion for 35s is determined at $X_a = 0.824$.

Code 11.3. Sizing a plug flow reactor (to be used with the function trapez in Code 11.2).

```
Function PF(Lower, Upper, tau, n)
Dim M, tol, c1, c2, c3, k
    c1 = 25 * trapez(0, Lower, n)-tau
    c2 = 25 * trapez(0, Upper, n)-tau
    M = Lower-c1 * (Upper-Lower)/(c2-c1)
    c3 = 25 * trapez(0, M, n)-tau
tol = 10 ^ -7: k = 0
While Abs(c3) > tol And k < 1000
If c1 * c3 < 0 Then
        Upper = M: c2 = 25 * trapez(0, Upper, n)-tau
    Else
        Lower = M: c1 = 25 * trapez(0, Lower, n)-tau
End If
M = Lower + (Upper-Lower) * c1 / (c1-c2)
c3 = 25 * trapez(0, M, n)-tau
k = k + 1
Wend
PF = M
End Function
```

As in other chapters, the integration can be applied to functions that require complicated numerical evaluations. An example is calculating the total heat produced in the reactors (Chapter 10).

12 Network Analysis

12.1 INTRODUCTION

Network analysis problems are a group of problems dealing with nodes and connections. The problems include the set of nodes and the "weight" factor for each connection between two particular nodes. In geographic terms, the nodes can represent locations, and weights are the distance. Nowadays, the problems seem too simple with Google Maps and Apple Maps. Yet network problems can be found in more complex applications besides finding pathways in a city. For example, the weight factor can be either time or cost in addition to distance. Of course, this kind of optimisation has puzzled many scientists over the centuries. Perhaps the most celebrated one is Leonard Euler's "The Seven Bridges of Königsberg".

While the network is graphically presented in a map form, the network data (connections and weight factors) can be easily stored in an Excel spreadsheet. Consequently, VBA code can be routinely employed.

12.2 CONVERTING A NETWORK TO TABLE FORM

The first step in the network analysis is to convert the 2-dimensional map (such as Figure 12.1) into a table form (Table 12.1). It is important to note that the nodes should have a numerical index for analysis. In the sample network, the node names (A, B, C, . . .) can be converted into numbers. However, for complex texts, such as location names, assigning a number to each node at the beginning might be more appropriate.

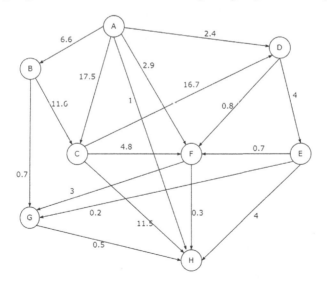

FIGURE 12.1 A network of eight nodes with directional connections.

DOI: 10.1201/9781003364511-12

TABLE 12.1

Distance between Points (the Empty Cells Indicating No Connection)

	A	B	C	D	E	F	G	H	J
1		1	2	3	4	5	6	7	8
2	1		6.6	17.5	2.4		2.9		1
3	2			11.6				0.7	
4	3				16.7		4.8		11.5
5	4					4	0.8		
6	5						0.7	0.2	4
7	6							3	0.2
8	7								0.5
9	8								

Please note that the cells represent one-way travel from row to column. Since all connections are dimensional, the other direction is zero (for example, from B to A).

12.3   ADVANCED EXAMPLE: SHORTEST PATHWAY PROBLEM

The most common problem is finding the shortest pathway between the two nodes in a network (Golden 1976). Amongst these methods, Dijkstra's method is one of the simplest and most effective. The method creates two sets of nodes: SSet (source nodes) and FSet (the remaining nodes). Initially, SSet contains a single node, the starting point (SN). Subsequently, the algorithm includes repetitive iterations, within which one node is moved from FSet to SSet. Finally, the iteration stops when the destination node (DN) moves into SSet (Foulds 1981).

Within the iterations, there are several calculation steps:

1. Each node in SSet has a cost value $d(i)$. Initial, SSet contains only SN, and $d(SN) = 0$. If node i is not in Sset, $Cost(i) = infinity$.
2. For each node in FSet, the distances to each node in SSet are obtained. These values are $c(i,j)$, where i is the node in SSet, and j is the node number in FSet.
3. The minimum of $d(i) + c(i,j)$ is calculated for node j ($MinCost(j)$). The corresponding node in SSet is recorded as $PreNode(j)$.
4. The minimum is determined from all nodes in FSet. Then, the corresponding node, $jmin$, is moved from Fset to Sset.
5. The process stops when $jmin$ is DN.

Once the iteration stops, the shortest pathway and corresponding sequence of nodes are output.

EXAMPLE 12.1.   THE SHORTEST PATHWAY BETWEEN TWO SPECIFIC NODES IN A NETWORK.

PROBLEM STATEMENT.

Find the shortest pathway between points A (column 1) and C (column 3) in Table 12.1. All connections are two way, and the weights are the same for both directions.

Solution

First, the input matrix must be checked and filled with two-way connections. The concept of "infinity" is not available in numerical coding. As a result, a huge but finite number can be used. The sum of all elements is used to ensure that this number is larger than all numbers in the matrix (Code 12.1). Here, the Excel function **sum** can be used within VBA.

It should be noted that VBA has a conversion function for alphabet and number (Chr and Asc). However, such a function can be worked within a small range. In general, one may need another code section to convert the node names into integers.

Code 12.1. Shortest pathway, part 1: setting up matrix M, filling out the bottom half and all empty cells (no link means infinity distance).

```
Option Base 1
Function shortestpath(A, SN, DN)
'SN is starting node, DS is the destination node
Dim M(), nNode, shortestNNode, N, infinity, i, j, k,
best_j
Dim temp As Variant
Dim Sset(), Fset() As Integer
Dim FCostNode As Variant
Dim MinCost, Prenode, Output, Cost, BestPath As Variant

'--check square matrix
nNode = A.Rows.Count
If A.Columns.Count <> nNode Then
    shortestpath = "Not a square matrix"
Exit Function
End If

infinity = 2 * Application.WorksheetFunction.Sum(A)
ReDim M(1 To nNode, 1 To nNode)
For i = 1 To nNode
    For j = i + 1 To nNode
```

```
          If A(i, j) > 0 Then
          M(i, j) = A(i, j)
     Else
          M(i, j) = infinity
     End If
          M(j, i) = M(i, j)
     Next j
Next i
```

The next section of the code defines the nodes in Sset and Fset. In this case, two arrays are used (SSet and Fset). Initially, all nodes are in Fset, so Fset(i) = i and Sset(i) = 0. If a particular node is moved from Fset to Sset, the values of the node in Fset and Sset are interchanged, Sset(i) = i and Fset(i) = 0.

Code 12.2. Shortest pathway, section 2: sizing SSet and FSet and initial values

```
ReDim Sset(1 To nNode): ReDim Fset(1 To nNode)
ReDim FCostNode(1 To nNode): ReDim Cost(1 To nNode)
ReDim MinCost(1 To nNode): ReDim Prenode(1 To nNode)
ReDim BestPath(1 To nNode)

For i = 1 To nNode
    Fset(i) = i
    Cost(i) = infinity
Next i
Sset(SN) = SN: Fset(SN) = 0: Cost(SN) = 0
```

In the second section of the code, the matrix sizing is required for the For Next loop. Initially, all nodes are in Fset with a cost value of infinity. Then node SN is moved to Sset.

Code 12.3. Shortest pathway, part 3: finding the shortest distance.

```
k = 1: shortestNNode = 1
Do While k < nNode
    temp = infinity
    For j = 1 To nNode
        If Fset(j) > 0 Then 'only calculate if j is Fset
            MinCost(j) = infinity
            For i = 1 To nNode
                If (Sset(i) > 0) And
(MinCost(j) > (Cost(i) + M(i, j))) Then
```

```
                                        MinCost(j) = (Cost(i) +
 M(i, j))
                                        Prenode(j) = i
                                        End If
                      Next i 'get the shortest cost to
connect j to a node in S- value is stored at
Prenode and MinCOst
                      If temp > MinCost(j) Then
                            temp = MinCost(j)
                            best _ j = j 'j moves to the
next value, and may not change best _ j
                      End If
                'reset cost j for next iteration
                End If
        Next j 'after j reaches Nnode, best_j is
identified
'-------move target j to Sset
        Sset(best _ j) = best _ j
        Fset(best _ j) = 0
        Cost(best _ j) = MinCost(best _ j)
    For j = 1 To nNode
        If Fset(j) > 0 Then
        Cost(j) = infinity
        End If
    Next j
    If best _ j = DN Then
        k = nNode
    Else
        k = k + 1
    End If
Loop
```

In the third section of the code, the best node **best_j** (gives the shortest distance to SN) is found and moved from Fset to Sset. The step is repeated until the best node is DN.

Code 12.4. Shortest pathway, part 4: finding the node on the shortest path backward.

```
'find nodes on the shortest path
BestPath(nNode) = DN:
k = nNode
Do While k > 0
BestPath(k-1) = Prenode(BestPath(k))
'Output(2 + nNode-k, 1) = BestPath(k-1)
```

```
        If BestPath(k-1) = SN Then
             shortestNNode = (nNode-k + 1)
             k = 0
        Else
             k = k-1
        End If
Loop
```

Since the best path from DN to SN is unknown (the path may not include all of the nodes in SSet), code is needed to find the nodes and sequence of the shortest path (Code 12.4). Consequently, the output is exported in Code 12.5.

Code 12.5. Shortest pathway, part 5: exporting the shortest pathway (can use "set cells" to improve visibility)

```
ReDim Output(1 To shortestNNode + 6, 1 To 1)
Output(1, 1) = "Shortest Distance"
Output(2, 1) = Cost(DN)
Output(3, 1) = "Number of nodes in the shortest path"
Output(4, 1) = shortestNNode
Output(5, 1) = "Nodes on the shortest path from S to D"
Output(6, 1) = SN

For i = 1 To (shortestNNode)
Output(i + 6, 1) = BestPath(nNode-shortestNNode + i)
Next i

shortestpath = Output
End Function
```

The answers to the problem are given in Table 12.2.

TABLE 12.2
Answers to the Shortest Distance

Shortest Distance

6

Number of nodes in the shortest path

4

Nodes on the shortest path from S to D

1

8

6

3

12.4   ADVANCED EXAMPLE: TRAVELLING-SALESMAN PROBLEM

The mathematical travelling salesman is one of the classical network problems. The problem is finding the shortest route to pass all nodes (once each) and return to the starting point (Conley 1988). The critical difference between this and the shortest path problem is that the pathway has to pass all points on the network. The starting point is not required. In general cases, this problem may not have a deterministic answer. The optimisation scheme is complicated. However, we can try Monte Carlo simulations with randomised numbers. With many trials, the code can get a good solution.

EXAMPLE 12.2.   SHORTEST PATHWAY COVERING ALL NODES IN A NETWORK.

PROBLEM STATEMENT

An engineer needs to check all listed devices in a processing plant. The time to move from one piece of equipment to another is listed in Table 12.1. Find a fast sequence of inspection routines for all devices.

Solution

In the beginning, all nodes are included in SSet. Subsequently, node 1 is added to the travel path and removed from SSet (SSet = 1). After that, the code randomly adds the remaining nodes to the travel path. In this case, the code needs to generate a random number for $N - 1$ times. Each time, a corresponding node from SSet will be added to the travel path and removed from SSet. At the end of the trial, the total distance is compared with *distancemin*. If the trialled distance is smaller than *distancemin*, then the distance becomes *distancemin* and is used to benchmark the subsequent trial. Otherwise, *distancemin* remains the same.

Code 12.6. Random search for a short pathway covering all nodes of a network.

```
Function shortesttravel(A, kmax)
Dim M(), nl, sel, distance, distancemin, nNode, k _
sol, N, infinity, i, j, k
Dim Sset(), Pathnode() As Integer
Dim Sol() As Variant
'-------
'using Code 12-1- setting up matrix M, filling out
the bottom half and all empty cells
'------
k = 1: distancemin = infinity * nNode
ReDim Sol(1 To nNode + 2)
```

```
While k < kmax
ReDim Pathnode(1 To nNode): ReDim Sset(1 To nNode)
Pathnode(1) = 1: Sset(1) = 1: distance = 0:
nl = nNode:
For j = 2 To nNode
nl = nl-1: sel = Round(Rnd() * (nl-1), 0) + 1:
    For i = 1 To nNode
            If Sset(i) = 0 And sel > 0 Then
            sel = sel-1
    End If
    If sel = 0 Then
    Sset(i) = 1: Pathnode(j) = i: sel = nNode
    distance = distance + M(Pathnode(j-1), Pathnode(j))
    End If
    Next i
Next j
If distancemin >= distance Then
distancemin = distance: k _ sol = k
    For i = 1 To nNode
    Sol(i) = Pathnode(i)
    Next i
End If
k = k + 1
Wend
Sol(nNode + 1) = distancemin: Sol(nNode + 2) = k _ sol
shortesttravel = Sol
End Function
```

The code results in the sequence of nodes, the obtained shortest distance and the trial number that produced this answer. In case multiple trials obtain the same answer, the last trial is shown. In the code, *kmax* is the maximum number of trials. With a reasonable *kmax* (around 20000) and some luck, the code can have the shortest cost of 18.5 with a corresponding sequence of {1,8,6,4,5,7,2,3}. One can also change the last line to "shortesttravel = M" to see if a correct matrix *M* is produced.

12.5   EXAMPLE FOR SELF-PRACTICE

EXAMPLE 12.3.   THE SHORTEST PATHWAY COVERS ALL NODES AND CONTAINS A PARTICULAR CONNECTION.

For a given network,

1 Find the shortest connection between any two points.
2 Find the shortest pathway that contains the previous connection.

Answer

Modify Code 12.6 to:

1 Find the shortest distance and the corresponding two points.
2 Use the starting point and run a Monte Carlo simulation from the second point.

Note that this solution is not always better than the solution from Code 12.6.

Appendices

List of Useful VBA Functions Which Return a Numerical Value

ABS	Returns the absolute value of a number
ASC	Returns an integer value representing a character code corresponding to the function's first character of a supplied string
ATN	Returns the arctangent of a number
COS	Returns the cosine of an angle
EXP	Returns e raised to the nth power
FIX	Returns the integer portion of a number
FormatNumber	Returns an expression formatted as a number
INT	Returns the integer portion of a number
LOG	Returns the natural logarithm of a number
RANDOMIZE	Used to change the seed value used by the random number generator for the RND function
RND	Generate a random number between 0 and 1
ROUND	Returns a number rounded to a specified number of digits
SGN	Returns the sign of a number
SIN	Returns the sine of an angle
SQR	Returns the square root of a number
TAN	Returns the tangent of an angle
MOD	Returns the remainder after division operator (Integer)
XOR	The bitwise exclusion operator

Other Useful VBA Functions

Array	Creates an array, containing a supplied set of values
Filter	Returns a subset of a supplied string array based on supplied criteria
ISEMPTY	Used to check for blank cells or uninitialised variables
ISERROR	Used to check for error values
ISNULL	Used to check for a NULL value
ISNUMERIC	Used to check for a numeric value
IsArray	Tests if a supplied variable is an array
IsMissing	Tests if an optional argument to a procedure is missing
IsObject	Tests if a supplied variable represents an object variable
Join	Joins a number of substrings into a single string

LBound	Returns the lowest subscript for a dimension of an array
UBound	Returns the highest subscript for a dimension of an array
REDIM	Initialises and resizes a dynamic array
BEEP	Produces a single beep noise
CALL	Transfers control to a subroutine or function
ERASE	Reinitialises the elements of an array (numerical variables are set to zero)
GOTO	Transfers control to the subroutine indicated by the line label

List of Useful Worksheet Functions (Using the WorkSheetFunction Object in VBA)

Atan	Calculates the arc tangent of a given number
Acos	Similar to Asin, Atan
Acosh	Similar to Asinh, Atanh
Average	Returns the average (arithmetic mean) of the arguments
Averageif	Calculates the average of the numbers in a range that meet supplied criteria
Cosh	Similar to Sinh, Tanh
Count	Returns number of numerical values within an array, ignoring text and empty elements
Degrees	Converts radians into degrees
Even	Rounds numbers up to the next even integer. The EVEN function always rounds numbers away from zero, so positive numbers become larger and negative numbers become smaller (i.e. more negative)
Fact	Returns a factorial to a number
ISErr	Returns a Boolean value indicating whether an expression is an error value, except #N/A
IsError	Returns a Boolean value indicating whether an expression is an error value (#N/A, #VALUE!, #REF!, #DIV/0!, #NUM!, #NAME?, or #NULL!)
IsEven	Returns TRUE if number is even or FALSE if number is odd. ISEVEN will return the #VALUE error if a value is not numeric
IsLogical	Returns TRUE when a cell contains the logical values TRUE or FALSE, and returns FALSE for cells that contain any other value, including empty cells
IsNA	Returns TRUE when a cell contains the #N/A error and FALSE for any other value or any other error type
IsNonText	Returns TRUE when a cell contains any value except text. This includes numbers, dates, times, errors and formulas that do not return text
IsNumber	Returns TRUE when a cell contains a number and FALSE if not
IsOdd	Returns TRUE when a value is an odd number and FALSE when a value is an even number. ISODD will return the #VALUE error if a value is not numeric
IsRef	Returns TRUE when a cell contains a reference and FALSE if not. You can use the ISREF function to check for a reference in a formula
Ln	Returns the natural logarithm of a number

Log	Returns the logarithm of a number to the specified base
Log10	Returns the base-10 logarithm of a number
LogEst	Calculates an exponential curve that fits your data and returns an array of values that describes the curve
Max	Returns the largest numeric value in the data provided
Min	Returns the smallest numeric value in the data provided
Median	Returns the median (middle number) in the supplied set of data
Mround	Rounds a given number up or down to the specified multiple. (*See Example 4.7.*)
Odd	Returns the next odd integer after rounding a given number up. The ODD function always rounds numbers away from zero, so positive numbers become larger and negative numbers become smaller (i.e. more negative)
Pi()	Returns value of Pi
Power	Returns a number raised to a given power
RandBetween	Returns a random integer number between the two specified numbers
Radians	Converts degrees to radians
Round	Rounds a number to a specified number of digits. (*See Code 4.1.*)
RoundDown	Rounds a number down to a specified number of digits
RoundUp	Rounds a number up to a specified number of digits
Slope	Returns the slope of a regression line based on known y and x values
SQRT	Returns the square root of a positive number. SQRT returns an error if number is negative
Sum	Sums all elements of an array

References

Chabert, J.-L., 1999. Methods of false position, in: Chabert, J.-L. (Ed.), A History of Algorithms. Springer, Berlin, Heidelberg, pp. 83–112. https://doi.org/10.1007/978-3-642-18192-4_4

Chapra, S.C., Canale, R.P., 2010. Numerical Methods for Engineers, 6th ed. McGraw-Hill Higher Education, Boston, MA.

Conley, W., 1988. Travelling salesman problem solved with simulation techniques. International Journal of Systems Science 19, 2115–2122. https://doi.org/10.1080/00207728808964103

Felder, R.M., Rousseau, R.W., 2005. Elementary Principles of Chemical Processes, 3rd ed. Wiley, Hoboken, NJ.

Fogler, S., 2006. Elements of Chemical Reaction Engineering, 4th ed. Pearson Education, Upper Saddle River, NJ.

Foulds, L.R., 1981. A multi-commodity flow network design problem. Transportation Research Part B 15, 273–283. https://doi.org/10.1016/0191-2615(81)90013-8

Golden, B., 1976. Technical note—shortest-path algorithms: A comparison. Operations Research 24, 1164–1168. https://doi.org/10.1287/opre.24.6.1164

Herráez, J.V., Belda, R., 2006. Refractive indices, densities and excess molar volumes of monoalcohols + water. Journal of Solution Chemistry 35, 1315–1328. https://doi.org/10.1007/s10953-006-9059-4

Himmelblau, D.M., Riggs, J.B., 2012. Basic Principles and Calculations in Chemical Engineering, 9th ed. International Series in the Physical and Chemical Engineering Sciences. Pearson, Boston, MA.

Ingram, G., Zhang, T., 2010. An introduction to the harmony search algorithm. Advances in Process Systems Engineering, 301–335. https://doi.org/10.1142/9789814299213_0009

Le, T.N., Phan, C.M., Nguyen, A.V., Ang, H.M., 2012. An unusual synergistic adsorption of MIBC and CTAB mixtures at the air-water interface. Minerals Engineering 39. https://doi.org/10.1016/j.mineng.2012.06.003

Levenspiel, O., 1999. Chemical Reaction Engineering, 3rd ed. Wiley, New York.

Lindfield, G.R., Penny, J.E.T., 2019. Numerical Methods: Using MATLAB, 4th ed. Elsevier, London.

Nawrocki, D., 2013. VBA Programming Techniques in Operations Research. 26th International Conference on Technology in Collegiate Mathematics VBA, pp. 255–263.

Pillai, H.K., Bandyopadhyay, S., 2007. A rigorous targeting algorithm for resource allocation networks. Chemical Engineering Science 62, 6212–6221. https://doi.org/10.1016/j.ces.2007.07.031

Press, W.H., Teukolsky, S.A., 1992. Adaptive stepsize Runge–Kutta integration. Journal of Computational Physics 6, 188. https://doi.org/10.1063/1.4823060

Serth, R.W., Lestina, T.G., 2014. Process Heat Transfer—Principles, Applications and Rules of Thumb, 2nd ed. Elsevier, Oxford.

Shashi Menon, E., 2015. Fluid flow in pipes, in: Transmission Pipeline Calculations and Simulations Manual. Elsevier, Oxford, pp. 149–234. https://doi.org/10.1016/B978-1-85617-830-3.00005-5

Shibata, Y., Hyde, A., Asakuma, Y., Phan, C., 2018. Thermal response of a non-ionic surfactant layer at the water/oil interface during microwave heating. Colloids and Surfaces A: Physicochemical and Engineering Aspects 556, 127–133. https://doi.org/10.1016/j.colsurfa.2018.08.010

White, D. 2002. ChE 224 Process Analysis Study Guide. Curtin University, Bentley.

Index

Note: Page numbers in *italics* indicate a figure and page numbers in **bold** indicate a table on the corresponding page.